Sounding Human

NEW MATERIAL HISTORIES of MUSIC

a series edited by
James Q. Davies *and*
Nicholas Mathew

ALSO PUBLISHED IN THE SERIES:

Sex, Death, and Minuets: Anna Magdalena Bach and Her Musical Notebooks
David Yearsley

The Voice as Something More: Essays toward Materiality
Edited by Martha Feldman and Judith T. Zeitlin

Listening to China: Sound and the Sino-Western Encounter, 1770–1839
Thomas Irvine

The Search for Medieval Music in Africa and Germany, 1891–1961: Scholars, Singers, Missionaries
Anna Maria Busse Berger

An Unnatural Attitude: Phenomenology in Weimar Musical Thought
Benjamin Steege

Mozart and the Mediation of Childhood
Adeline Mueller

Musical Migration and Imperial New York: Early Cold War Scenes
Brigid Cohen

The Haydn Economy: Music, Aesthetics, and Commerce in the Late Eighteenth Century
Nicholas Mathew

Tuning the World: The Rise of 440 Hertz in Music, Science, and Politics, 1859–1955
Fanny Gribenski

Music in the Flesh: An Early Modern Musical Physiology
Bettina Varwig

Creatures of the Air: Music, Atlantic Spirits, Breath, 1817–1913
J. Q. Davies

Sounding Human

MUSIC AND MACHINES,
1740/2020

Deirdre Loughridge

THE UNIVERSITY OF CHICAGO PRESS
CHICAGO AND LONDON

The University of Chicago Press, Chicago 60637
The University of Chicago Press, Ltd., London
© 2023 by The University of Chicago
All rights reserved. No part of this book may be used or reproduced in any manner whatsoever without written permission, except in the case of brief quotations in critical articles and reviews. For more information, contact the University of Chicago Press, 1427 E. 60th St., Chicago, IL 60637.
Published 2023
Printed in the United States of America

32 31 30 29 28 27 26 25 24 23   1 2 3 4 5

ISBN-13: 978-0-226-83009-4 (cloth)
ISBN-13: 978-0-226-83011-7 (paper)
ISBN-13: 978-0-226-83010-0 (e-book)
DOI: https://doi.org/10.7208/chicago/9780226830100.001.0001

Library of Congress Cataloging-in-Publication Data

Names: Loughridge, Deirdre, author.
Title: Sounding human : music and machines, 1740/2020 / Deirdre Loughridge.
Other titles: New material histories of music.
Description: Chicago : The University of Chicago Press, 2023. | Series: New material histories of music | Includes bibliographical references and index.
Identifiers: LCCN 2023016766 | ISBN 9780226830094 (cloth) | ISBN 9780226830117 (paperback) | ISBN 9780226830100 (e-book)
Subjects: LCSH: Mechanical musical instruments—History. | Music and technology—History. | Electronic music—History and criticism. | Music—Performance—Philosophy and aesthetics. | Popular music—Production and direction. | Auto-tune (Computer file)
Classification: LCC ML1050.L68 2023 | DDC 786—dc23/eng/20230501
LC record available at https://lccn.loc.gov/2023016766

*To my loving parents,
Marianne and Michael Kelly*

Contents

List of Audio Examples ix

INTRODUCTION:
SOUNDING HUMAN WITH MACHINES 1

1: BECOMING ANDROID:
REINTERPRETING THE AUTOMATON FLUTE PLAYER 19

2: HYBRIDS:
VOICE & RESONANCE 51

3: ANALOGIES:
DIDEROT'S HARPSICHORD & ORAM'S MACHINE 79

4: PERSONIFICATIONS:
PIANO DEATH & LIFE 113

5: GENRES OF BEING POSTHUMAN:
CHOPPED & PITCHED 133

CODA: LEARNING MACHINES 159

Acknowledgments 177
Notes 181
Works Cited 219
Index 233

Audio Examples

This list of recordings corresponds to the audio examples referenced throughout the book. As of the time of publication, they are available on a Spotify playlist, a link to which is provided at https://press.uchicago.edu/sites/loughridge/index.html.

1: Wolfgang Amadeus Mozart, Fantasie in F Minor, K. 608 ("Ein Orgelstück für eine Uhr"), performed by Jean-Pierre Lecaudey, *Mozart: Les 17 sonates d'église & L'oeuvre pour orgue* (2006) 5

2: Kraftwerk, "The Robots," *The Man-Machine* (1978) 9

3: Afrika Bambaataa and The Soulsonic Force, "Looking for the Perfect Beat," *Planet Rock: The Album* (1986) 10

4: Nona Hendryx, "Transformation," *Nona* (1983) 11

5: Giovanni Battista Pergolesi, *La serva padrona*, final duet "Per te ho io nel core," performed by Gilbert Bezzina, Isabelle Poulenard, and Ensemble Baroque de Nice, *Pergolesi: La serva padrona* (2010) 15

6: Wolfgang Amadeus Mozart, *Die Zauberflöte*, K. 620, act 2, scene 3, "Der Hölle Rache," performed by Cyndia Sieden and English Baroque Soloists, directed by John Eliot Gardiner, *Mozart: Die Zauberflöte* (1996) 15

7: Jacques Offenbach, *Les Contes d'Hoffmann*, act 2, "Les oiseaux dans la charmille," performed by Dame Joan Sutherland and Orchestre de la Suisse Romande, *Offenbach: Les Contes d'Hoffmann* (1972) 15

8: Wolfgang Amadeus Mozart, *Die Zauberflöte*, act 1, scene 3, "Das klinget so herrlich, das klinget so schön," directed by John Eliot Gardiner, *Mozart: Die Zauberflöte* (1996) 15

9: François Francoeur, *Pirame & Thisbé*, prologue, performed by Choeur de l'Académie Baroque, *Pirame & Thisbé* (2008), from 00:42 35

10: Wolfgang Amadeus Mozart, *Die Zauberflöte*, act 2, scene 5, "Ein Mädchen oder Weibchen," performed by Gerald Finley and English Baroque Soloists, directed by John Eliot Gardiner, *Mozart: Die Zauberflöte* (1996) 52

11: Wolfgang Amadeus Mozart, *Die Zauberflöte*, act 2, scene 9, "Klinget Glöckchen, klinget," directed by John Eliot Gardiner, *Mozart: Die Zauberflöte* (1996), from 4:35 52

12: Alessandro Striggio, arranged by Peter Philips, "Chi fara fede al Cielo," *Renaissance Music at the Court in Heidelberg* (1996) 57

13: Marin Marais, Prelude No. 4 from Suite No. 1 in D minor, *Marais: Pièces de viole du premier livre, Première partie, Suites No. 1, 2 & 4* (2006) 59

14: Marin Marais, "Fantasie luthée," from Suite No. 2 in D minor, *Pièces de viola*, vol. 2 (2001) 68

15: Marin Marais, "Les cloches ou carillon," from Suite No. 2 in D minor, performed by Markku Luoloajan-Mikkola (viola da gamba), *Marais: Pièces de viole du second livre* (1998) 68

16: Michele Mascitti, Violin Sonata in G Major, Op. 2 No. 5, mvmt. 3 (Corrente Andante), performed by Fabrizio Cipriani (violin) and Antonio Fantinuoli (cello), *Mascitti: 6 sonate da camera, Op. 2* (1997) 70

17: François Couperin, "Les Sentimens Sarabandes," from *Pièces de Clavecin*, book 1, performed by Kenneth Gilbert, *Couperin: Premier livre de clavecin* (1971) 88

18: Daphne Oram, "Four Aspects," *An Anthology of Noise and Electronic Music*, vol. 2 (2006) 94

19: Daphne Oram, "Contrasts Essconic," *Endless Waves: The Dawn of Electronic Noise & Ambient Music*, vol. 2 (2021) 99

20: Calvin Harris and Rihanna, "This is What You Came For" (2016) 133

21. Kiiara, "Gold," *low kii savage* (2016) 133

22: Zara Larsson and MNEK, "Never Forget You," *So Good* (2017) 133

23: T-Pain featuring Ludacris, "Chopped N Skrewed," *Thr33 Ringz* (2008) 139

24: Cher, "Believe," *Believe* (1998) 145

25: Flume, "Never Be Like You," *Skin* (2016) 147
26: MNEK, "At Night (I Think About You)" (2016) 155
27: David Cope, "From Darkness, Light," Prelude, performed by Mary Jane Cope and Erika Arul, *Emily Howell: From Darkness, Light* (2010) 159
28: Laetitia Sonami, "Breathing in Birds and Others," *Metagesture Music (Atau Tanaka Presents)* (2017) 162
29: Dadabots, "We Generated This Album in Our Sleep," *Can't Play Instruments*, vol. 1 (2021) 166
30: Dadabots, "We Haven't Listened to This Track Yet," *Can't Play Instruments*, vol. 1 (2021) 166
31: Holly Herndon, Jlin, Spawn, "Godmother," *PROTO* (2019) 168
32: Jlin, "Expand," *Dark Energy* (2015) 168
33: Jlin, "Embryo," *Embryo* (2021) 168
34: Arca, "Riquiquí," *KiCk i* (2020) 172
35: Arca, "Riquiquí(xxv25)," *Riquiqui; Bronze-Instances(1–100)* (2020) 172
36: Arca, "Riquiquí(xxxiii33)," *Riquiqui; Bronze-Instances(1–100)* (2020) 172
37: Arca, "Riquiquí(lxxvii77)," *Riquiqui; Bronze-Instances(1–100)* (2020) 172
38: Arca, "Nonbinary," *KiCk i* (2020) 173

INTRODUCTION

Sounding Human with Machines

Let's begin with a game. There are four pieces of music, and according to the rules of the game, at least one of them was composed by Frédéric Chopin and at least one by a computer program. The player's goal is to identify which pieces were created by which entity. Composer David Cope set up this game in his 2001 book *Virtual Music*, explaining that it is about "recognizing human-composed music as distinct from machine-composed music."[1] Typically, players' answers reveal their inability to tell which is which. For some, this result is not only surprising but also deeply disturbing. The reaction of cognitive scientist Douglas Hofstadter is exemplary: "I was truly shaken. How could emotional music be coming out of a program that had never heard a note, never lived a moment of life, never had any emotions whatsoever?"[2] For Hofstadter, the idea of computer-composed music was incompatible with music's grounding in human experience and being—a grounding he summed up by describing pieces of music as "soul-to-soul messages."[3] Reflecting on reactions of this type, Cope observed, "I'm messing with some pretty powerful relationships here and doing so in a mechanical way. If I had done it myself, as a human being, these individuals could probably live with it.... But somehow using HAL"—the computer intelligence in the Stanley Kubrick–directed film *2001: A Space Odyssey* that decides to eliminate its human companions—"or some version of HAL is the ultimate insult."[4] For Cope, players' failures at the game confirmed his success at getting a computer program to generate "interesting" and "convincing" music.[5]

Another implication to Cope's game, however, is rooted less in its outcome than in its structure. To participate in the game, one must assent to the premise that there is "human-composed music as distinct from machine-composed music." In other words, the game requires a binary choice: either human or nonhuman is responsible for the music. But we could instead think of the computer program as human created and listen to its music as jointly human-machine composed. We could also think about how machines were involved in the human-composed music. In the case of Chopin, we might hear how his compositions were shaped by the pianos on which he played, recognizing the piano as a machine with a particular set of affordances and constraints in relation to the human body.[6] To listen in these ways, for music as a joint product of human and nonhuman beings, with creative contributions distributed among them, is to break Cope's game.

This book is about the making and breaking of the logic of Cope's game. It is about how categories of human and machine have been produced and experienced in relation to one another through music making and listening, and it is about how musical artifacts have been—or can be—used to help explain and contest what it is to be human. There is a stability to the "human *or* machine" logic in musical discourse and practice stretching from around the 1750s to the early twenty-first century. It was in 1754 that court flute player and composer Joachim Quantz proposed that although an android flute player might astonish with superhuman rapidity and precision of performance, "it would never move you."[7] While the sonic, social, and technological particulars have changed, the structure of Quantz's opposition remains recognizable in experiences like Hofstadter's, the construction *it would never* defining a boundary of absolute difference between human and machine begging to be tested and traversed with world-destabilizing, anxiety-provoking effects. Across the nineteenth and twentieth centuries, recurrent practices of parsing "human" musicality from its "merely mechanical" substrates or simulations appear in fictional narratives about musical androids, pedagogical literatures on musical performance, racist and sexist denials of musicians' creative agency and musical understanding, and the reception of various music-technological innovations.[8]

The either-or logic is even embedded in a popular definition of music: ethnomusicologist John Blacking's notion of music as "humanly organized sound." To clarify the human involvement required for sounds to become music, Blacking turned to boundary cases involving machines. "The regular beats of an engine or pump" would be excluded from music, he explained, "because their order is not directly produced by human beings." The sound of a Moog synthesizer, on the other hand, would be included "as long as it was only the timbre and not the method of ordering that was outside human control."[9] While Blacking attributed these categorical judgments and rationales to the Venda community whose music he studied, he also invested them with a universal validity. One senses Blacking's own anxiety about control shifting from humans to machines.

He was far from alone in manifesting this anxiety during the 1970s, when his study was published and when synthesizers and computers were playing new roles in organizing sound. Beyond the common concerns about loss of expressivity or musicianship, Blacking had a specific intellectual stake in direct human control over the temporal ordering of sound: such control was necessary to ground the kind of sociocultural analysis of music he sought to do, which would disclose relationships between the "patterns of human organization and the patterns of sound produced as a result of human interaction."[10]

Today, in the early twenty-first century, I find myself, along with many other musicians and scholars, in an era of unmaking the "human *or* machine" logic and seeking out others better characterized by conjunctions such as *and* or *with*. In part, the shift is a function of the changing qualities and capacities of the machines around us. In 1960, psychologist and computer scientist J. C. R. Licklider differentiated three types of "man-machine systems." His first two categories described the familiar: in "mechanically extended man," mechanical parts of the system extend a body part such as an arm or eye, but all initiative and direction comes from the human; in "humanly extended machines," the goal is full automaticity, but some rudimentary input from a human operator remains necessary. The third type was a novel model, a dawning possibility: in "man-computer symbiosis," there would be a "cooperative 'living together'" in intimate association, the computer contributing in ways normally associated with intelligence.[11] When composer-performer-programmer George Lewis describes the "values such as relative autonomy, apparent subjectivity, and musical uniqueness rather than repeatability" he has built into the interactive computer system with which he and others improvise, he is describing a realization of the third type of relationship.[12] Indeed, according to fellow composer and designer of interactive computer systems for musical performance Joel Chadabe, "these instruments introduced the concept of shared symbiotic control of a musical process."[13]

And yet, changing capabilities are not enough to explain transformed conceptions of machines or social imaginaries of their present and future. To begin, one might wonder why it was clear to Licklider and Lewis that computers represented new forms of machine calling for new kinds of human-machine relationship and why others greeted these unfamiliar entities with familiar "human *or* machine" anxieties directed toward machines encroaching on uniquely human territory and threatening total replacement. The questions become, as anthropologist Lucy Suchman has written, "how and when the categories of human or machine become relevant, how relations of sameness or difference between them are enacted on particular occasions, and with what discursive and material consequences."[14] To approach the making and breaking of the either-or logic with these questions in mind is to say that we are dealing not with an inevitable history and future of machine advancement toward symbiosis or replacement, but rather with a history and future of negotiations between the abstract categories *human* and *machine* and the material specificities of particular situations. About these situations, we can also inquire, with Donna Haraway, "which kinds of humanness and machineness are produced out of those sorts of material-semiotic relationships."[15]

My interest in this book is with the edges of the resilient *human or machine* formation that made music a defining feature of *the human* on condition of being *not machine*. Its focus is on listeners who had proximity to the crystallization of this logic in the 1750s but who experienced and made sense of human-machine relations in other ways; with unorthodox thinkers who used musical artifacts to develop materialist theories of human capacities and who regarded sound as substantiating human-machine similarity more so than difference; and with twenty-first-century contexts where musical instruments and sounds defy categorization as animate or inanimate, with or without soul, human or machine, calling for alternative modes of listening and understanding. In early Enlightenment France, human-machine opposition was not a default mode of musical thought, but the material conditions and ideas that would go into making it so were coming together. In early twenty-first-century America, musical encounters prompted reexaminations of intuitions around the categories of human and machine, denaturalizing prior assumptions and familiarizing alternate conceptions and relationships. Focusing on edge cases is conducive to discerning the processes involved in the movement between abstract and particular, conceptual and material versions of human and machine—the interactions between familiar and novel experiences that sometimes reproduce existing logics and sometimes generate new ones.

This study thus takes generative uncertainty about what defines or characterizes human and machine both as subject matter and as method, with Suchman's questions of *how*, *when*, and *with what consequences* in mind. Recent scholarship has done much to unsettle naturalized assumptions about machines, revealing how not only their paradigmatic instances but also their figurative meanings, cultural status, and affective charge have taken different forms at different times. In large measure, this has meant dislodging industrial-age anxieties about machines and assumptions about their limitations ("anything pretending to be art cannot come out of a machine," Jacques Barzun aptly summed up the conventional wisdom in 1961) in order to recover historically specific understandings.[16] Annette Richards homes in on this problem, for example, in a study of Mozart's F minor Fantasie K. 608. Mozart composed this twelve-minute work, which comprises a set of variations framed by fugal sections, for the mechanical organ of a musical clock. Richards shows the sense of incongruity between musical genius and machine "so central to modern readings" of the work to be absent from its early reception. Instead, she finds a late eighteenth-century enthusiasm for mechanical displays and a sense of their compatibility with the creative faculty of a rational soul, suggesting we hear K. 608 not as musical despite its mechanical medium but rather as a work of "superhuman virtuosity... [that] celebrates the wonderful mechanical potential of the organ clock" (audio 1).[17] Examining epistemological and practical developments around instruments and machines in the early modern period, Bonnie Gordon and Rebecca Cypess have demonstrated connections between their newly positive status in science (natural magic) and music. Thus, Gordon reinterprets Monteverdi's *L'Orfeo* by highlighting moments that demonstrate Orfeo's reliance on his lyre, finding Monteverdi's musical dramatization to perform not the superiority of expressive song over vocal virtuosity (as modern interpreters have tended to think), but rather the power of technology and its extension of human capacities.[18]

Such newfound interest in recovering positive, collaborative conceptions of machines in music history reflects the waning of the *human or machine* paradigm that made them oppositional terms. Indeed, I would account for my own interest in the history of how human and machine have been understood in relation to one another in this way. "By the late twentieth century," Haraway declared in 1985, "we are all... theorized and fabricated hybrids of machine and organism; in short, we are cyborgs."[19] *Posthuman* is one name for this condition. N. Katherine Hayles offered an influential account of "how we became posthuman" in her 1999 book of that name focusing on the development of cybernetic science and science fiction literature since World War II.[20] A year earlier, Kodwo Eshun identified a "posthumanization which used to be called dehumanization" in the work of musicians who joined themselves to machines and synthetic processes, such as Kraftwerk and Roger Troutman's Zapp.[21] According to Eshun, he and fellow cybertheorists in the mid-1990s "got a particular boost from music": electronic dance genres like drum 'n' bass "left the song far behind," creating a sonic experience that "obliged you to come up with a conceptual apparatus which was totally post-human."[22] From underground and avant-garde, posthuman musical configurations increasingly moved into the mainstream of popular music, enabling Alexander

Weheliye and Joseph Auner to discuss "posthuman voices" in contemporary pop music in the early 2000s.[23]

What does the eighteenth century have to do with these developments? As Roger Grant observes, "it is a trope of posthumanist scholarship to call up Enlightenment culture in order quickly to dismiss it, arguing that the eighteenth century amplified and consolidated the humanism it had inherited from the Renaissance."[24] The *Enlightenment subject* has often served as a synonym for the version of the human that posthumanism seeks to critique and replace. However, scholars have gathered mounting evidence that Renaissance and Enlightenment cultures were more putatively posthumanist than humanist in their assumptions when it comes to thinking about machines in relation to music and *the human*.[25] To date, little has been made of this apparent discrepancy. The disciplinary norms of musicology are such that scholars typically focus on a particular repertoire, historical period, and geographic region, gaining valuable depth in that area but limiting the opportunities to pursue comparisons or connections across time and place.[26]

If the early twenty-first century is a time of mutually exclusive "human or machine" configurations losing sway, there are compelling reasons, I contend, to look to the eighteenth century for a mirror image process. In their history of the enlightenment, Clifford Siskin and William Warner argue that between the writing of natural philosopher Francis Bacon's *Great Instauration: The New Organon* (1620) and critical philosopher Immanuel Kant's *What Is Enlightenment?* (1784), a fundamental change took place that "is nowhere more evident than in the ways they used the word 'machine.'"[27] For Bacon, machines were essential to man's ability to advance knowledge, the stalled progress of which was due to men's foolish efforts to use their intellects alone. "Neither the bare hand nor the unaided intellect has much power; the work is done by tools and assistance, and the intellect needs them as much as the hand," Bacon noted.[28] "The mind should not be left to itself, but be constantly controlled; and the business done (if I may put it this way) by machines."[29] Kant, by contrast, urged men to "use your own reason" and called for a freedom of public discourse by which they would exercise their independent thought and become "more than machines."[30] *Machines*, for Kant, represented merely obedient objects and described the state of men in societies that curtailed the free use of reason, embodying the very antithesis of what men should be. Zeroing in on the history of the automaton as a conceptual object, Minsoo Kang observes that in the late seventeenth and early eighteenth centuries, "likening a person to a machine . . . carried the positive valence of an intricate, well-functioning, and beautiful device," whereas from the mid-eighteenth century onward, the comparison became "overwhelmingly negative in character," indicating "a person who lacks the principle of freedom, stunted in thought and spirit through either external oppression or witless conformism."[31] The changed figurative uses of machine reflect, as Jessica Riskin suggests, "a newly configured landscape of machines and people."[32]

We might add that this newly configured landscape was also a soundscape. But more than a "scape" in which to snapshot old and new configurations, sound was involved in processes of reconfiguration. How so is suggested by two ways of reading the phrase *sounding human* of this book's title—as referring to ideas about what sounds human, held prior to what makes such sounds in any particular situation and portable across contexts, and as referring to empirical events, to humans making sounds. Attending to how people have made sense of sounds enables tracking interactions between these two readings of *sounding human*—between prior ideas and new encounters.

An extensive body of scholarship has examined the interactions between prior ideas and new encounters in the contexts of Europeans' incursions across the globe. Figures of the nonhuman loom large in the colonial archive, and from such sources scholars have endeavored to analyze not only how colonists' listening practices cast Others as bestial or diabolical, but also how such putatively subhuman sounds resonated differently with Indigenous epistemologies, ontologies, and cosmologies. As Ana María Ochoa Gautier argues with respect to the vocal practices of *zambo* (African-Amerindian) boat rowers in nineteenth-century Colombia, "If sounding like animals, learning sounds from animals, or incorporating nonhuman entities in sound is not a problem but an objective, then it becomes evident that the human-nonhuman relation, or the relation between nature and culture present in the voice is not one that debases the person."[33] Advancing Indigenous sound studies, Dylan Robinson diagnoses "the self-censoring listening of settler colonialism that avoids certain kinds of listening experience, and especially ones that would affirm human-nonhuman relationships."[34] To focus on human-machine relationships offers a particular refraction of these *longue durée* dynamics, another historical lens through which to render structuring categories contingent rather than self-evident and to understand them as outcomes of processes rather than as givens.

Engaging with the specificity of new encounters in the eighteenth century can yield some significant revisions to conventional wisdom, such as to thinking about musical androids. As theorized by Wilhelm Jentsch and Sigmund Freud in the early twentieth century, androids are paradigmatically uncanny, productive of that special disquiet stemming from uncertainty about their animate or inanimate status.[35] Throughout the nineteenth and twentieth centuries, writers of fact and fiction alike firmly associated androids with such troubling uncertainties, linking them to philosophical debates around materialism and to industrial contexts for the mechanization of labor and life. In 1970, roboticist Masahiro Mori introduced the idea of an "uncanny valley" to account for the unsettling effect of manufactured objects that approach human lifelikeness. Proposing that "we should begin to build an accurate map of the uncanny valley, so that through robotics research we can come to understand what makes us human," he implied such a map would be static, descriptive of degrees of lifelikeness inherent in objects.[36] In a detailed study of the making and reception of keyboard-playing women automata, however, Adelheid Voskuhl pries these artifacts apart from their later associations, elaborating a context in which they were appreciated as works of mechanical artisanship, modeling sociability and sentimental selfhood—and not immediately uncanny. Such findings support Laura Mulvey's observation, shown through the case of reactions to the moving images of cinema, that the "uncertainties particular to the human mind" productive of the uncanny are not unchanging but rather shaped by prior experiences and beliefs—"a 'technological uncanny' waxes and wanes."[37]

How the automaton flute player made by Jacques Vaucanson in 1737 could be an object of pleasure rather than anxiety when it was new and how its identity changed in the 1740s will be a subject of chapter 1. The point to be made here is that the uncanny is revealing of boundary lines that have been internalized in such a way that their confusion or destabilization becomes unsettling. The notion of a "human-machine boundary" is now so ubiquitous as to be thoroughly naturalized, and one is hard-pressed to find a discussion of musical machines that does not assume that such a boundary exists. But *boundary* is a very specific relation, entailing neighboring positions and violation when crossed. It is not self-evident that humans and machines should be thought of as adjacent to one another, separated by a boundary that they can traverse depending on their display of particular characteristics or capacities. Nor is it obvious that music, or certain musical qualities, should be placed on the human side of that boundary, such that their appearance on the other side is disturbing (or rather, that music should be something that readily traverses the boundary, but with a resilient power to unsettle whenever it does). Revisiting eighteenth-century sources without the assumption of a human-machine boundary makes room for how musical encounters were involved in *producing* that boundary as a specific relation.

This dynamic theory of the uncanny suggests that the prevalence of uncanny androids in the nineteenth and twentieth centuries provides a measure of the strength *human* and *machine* had acquired as predefined categories, each expected to be stable in itself and its difference from the other. Lynn Festa has argued that while late Enlightenment discourses deployed a self-evident human in service of the political project to secure rights in its name, literature and art of the early eighteenth century "often worked without a positive, stable concept of humanity as an organizing rubric." That is, in the first half of the eighteenth century, humanity was not a "known quantity" writers and artists could mimetically represent; rather, they "defined and produced" conceptions of the human through their works.[38] Studies of the nineteenth century, by contrast, show the ongoing processes of producing conceptions of the human to take place in relation to more firmly established "organizing rubrics," to use Festa's phrase, or "templates," to use Louis Chude Sokei's term. Chude Sokei traces how anxieties about technology were conflated with anxieties about race and empire. Proto-sci-fi narratives of machine uprising, for instance, betray an unmistakable basis in White anxieties about Black freedom and personhood (note the hardened racial binary). As preoccupations with technological and social change took hold in the nineteenth-century American context of industrialization and slavery, Chude Sokei argues, the conditions were set for metaphors of race and technology to be "deliberately made to coincide," above all in anthropomorphized machines, and for "race [to] provide the template for how the West frames its relationship with those nonhuman technological transformations."[39]

The vocoder illustrates how the perception of a sound has been shaped by this racialized template and used to contest it. By combining linguistic sounds from a voice with tones from an electronic source, the vocoder has offered a site for experiencing and interpreting relations between human and machine. Discussing electro artists like The Jonzun Crew, Eshun explained that "the vocoder turns the voice into a synthesizer," which "lets you... become cartoon, become animal, become supercomputer."[40] The favored perception of what the vocoder does to the voice, however, is to make it sound robotic.[41] Kraftwerk made this interpretation explicit, reciting "we are the robots" into a vocoder on their 1978 album *The Man-Machine* (audio 2). The track exemplifies what band member Florian Schneider explained as "expos[ing] the mechanical and robotic attitude of our civilization," in which "many people are robots without knowing it."[42] In an influential account of Afrofuturism, Tricia Rose proposed that "what Afrika Bambaataa and hip-hoppers like him saw in Kraftwerk's use of the robot was an understanding of themselves as *already having been robots*." Given their reduction to "labor for capitalism" with "very little value as people in this society" as legacies of slavery in the United States, these Black artists "master[ed] the wearing of this guise in order to use it *against* [their] interpolation" (audio 3).[43] Following the development of vocoder effects into 1990s R&B, Weheliye argued that their robotic sound, ironically, "lends an aura of increased 'humanity' and 'soulfulness' to the singer's voice," thanks to its nostalgic association with an earlier, more analog decade. He also interpreted the use of vocoder effects in R&B as a critique of historical narratives in which the posthuman replaces "the human" construed exclusively as "a white liberal subject." Weheliye thus accounted for how this sound in Black popular music works to reframe "the humanist subject... to include the subjectivity of those who have had no simple access to its Western, post-Enlightenment formulation."[44]

Overshadowed by these musical uses and discussions around the vocoder is another possible hearing of the human-machine relationship it sounds. In 1971, Wendy Carlos created an electronic version of the finale from Beethoven's Ninth Symphony using Moog synthesizers and a vocoder.[45] Recent discussions of this track share the perception of human voice turned robotic. Christine Lee Gengaro describes the sound of "voices—disconnected from their humanity—ironically singing about universal brotherhood."[46] For Judith Peraino, it is "an electronic sound that resembles the human voice, but with its humanity removed."[47] These are compelling interpretations, consistent with the experience of Carlos's substitution of electronically synthesized tones for the familiarly human sound of singing voices. And yet, contrast these accounts with the one offered by Carlos's record producer and musical collaborator Rachel Elkind (who performed the words for the track): "We played [the vocoded 'Ode to Joy'] to friends, and they were somehow terrified. Here was a machine trying to sound like a human, and getting so close. It was just like HAL all over again."[48] Elkind's account differs significantly in the directionality of its human-machine relationship: rather than the sound of voices minus humanity, it is the sound of a machine approaching humanity. This interpretation is consistent with the marketing of the vocoder, which portrayed it as imparting the powers of speech to electronic instruments. As one 1978 ad put it, the vocoder "makes any electronic sound source... appear to talk."[49] It is also consistent with the perspective of Carlos's creative process; as a composer who worked with synthesized sound, she was adding verbal sounds to her medium rather than taking away the humanity of voices. Carlos herself described figuring out how to make the vocoder "not just speak, but sing," and introducing listeners to the experience of a "'singing' synth."[50] The resulting terror, according to Elkind, springs from the imagined step toward a HAL scenario of an intelligent machine wresting control and turning on its humans. For an example of how else it can make a difference to hear the vocoder imparting speech to tones, consider Nona Hendryx's 1983 song "Transformation"; though the song has been described as "robotic," the vocoder effect is reserved for the first-person voice of—we might glean from Hendryx's lyrics—Mother Nature, figured as ruling the "variations, alternations, deviations" that are "us all."[51] To hear the vocoder as altering a human voice in "Transformation" is thus to miss how it works to personify the impersonal power of Mother Nature and to do so in a way that denies an opposition between nature and technology (audio 4).

I am interested in moments like the initial one around Carlos's "Ode to Joy" for the alternate perceptual and imaginative possibilities they open up and for what they reveal about why other forms of experience and sense-making prevailed. Such dynamics are often to be found around novelties, like Jacques Vaucanson's flute player (chap. 1), Antoine Ferrein's theory of the voice (chap. 2), Daphne Oram's sound wave instrument (chap. 3), and the sounds of Auto-Tuned and chopped and pitched vocals (chap. 5). They also arise at points of potential obsolescence, as with the viola da gamba (chap. 2) and acoustic pianos (chap. 4). Digging into such moments, into the specificities of musical encounters, leads to people and material objects that have had little presence in histories oriented toward relating musical works to canonical philosophical debates. Looking for surrounding context for bits of automaton reception quoted in scholarly literature, for instance, led me to Françoise de Graffigny's letters and the play *L'Oracle*, discussed in chapter 1. The existence of the Daphne Oram Collection—an extensive archive of the electronic composer-inventor's written records and tapes—turned reading her book *An Individual Note* into an opportunity to try to piece together sources, motivations, and conditions for her account of "the human" and how to "humanize our machines."[52] Such writings allow for more fully mapping the field of possibility on which musical encounters took place and the processes by which certain paths were taken—even as they also reveal the contingencies of material survival and access and of the histories thereby enabled.

Each chapter addresses a figure of relation between human and machine—android, hybrid, analogy, personification, and posthuman—through examples that move from eighteenth-century makings of "sounding human versus machine" to twenty-first century needs for alternative configurations. Chapter 1 revisits Vaucanson's flute player, a musical automaton that has come to be known as "the first actual android."⁵³ Centering the reception of androids as distressingly destabilizing of a human-machine boundary, however, opens a significant gap between the debut of Vaucanson's flute player and the arrival of androids into the sociocultural imaginary. Recovering the network of meanings within which the flute player was made and first received reveals how mythological symbolisms converged with colonial fantasies to shape encounters with the automaton and its "becoming android." The writings of Graffigny—a bestselling author of the French Enlightenment—illuminate how Enlightenment pleasure at automata was not just about appreciating the mechanical genius of their makers or delighting in a subversive materialism but was also a performance of one's own sensibility, a weighty skill in a social world that made sensibility evidence of humanity where humanity was uncertain.

Chapter 2 stays with ca. 1740 France and turns to the fascination with hybrids that paralleled the reception of musical automata before they became android. The lawyer and amateur musician Hubert Le Blanc wanted to defend the viol against the rising popularity of the violin, and the anatomist Antoine Ferrein wanted to explain the physiology of the voice. Despite their different aims, each devised a new taxonomic system for musical instruments based on sonic qualities and their relation to materiality; and each did so not to fit their object of interest into a discrete category but to better account for it as a hybrid. Their arguments reflect listening practices that transitioned between earlier conceptions of the voice and harmony as God-fashioned models for human music making and later conceptions of the voice as the sound of a pure, natural humanity. Le Blanc's text additionally suggests how the conceptual and perceptual techniques they brought to sound were also central to natural history and global empire, which placed high epistemological and social stakes on the production of hybrids. The forgotten status of viol and voice as hybrids, finally, exemplifies a historical pattern of hybrids—rather than undoing binaries—working to naturalize them, the hybrid disappearing while the distinct categories required to make it remain.

Chapter 3 turns from hybrids to analogies; where hybrids combine distinct categories into a blend, analogies create a relation of likeness across difference—in the cases here, between something readily explained and something difficult to explain to better understand the latter. That human-machine analogies have played significant roles in the histories of neuroscience, cognitive science, and computer science is well known. The omission of human-musical instrument analogies from these field's genealogies, however, attests to the contingency of the category of machine and the special ability of musical instruments to bring that contingency into view. Pairing Denis Diderot's mid-eighteenth-century use of the harpsichord and Oram's mid-twentieth-century use of electronics and her own invented "sound wave instrument" or "Oramics machine" for converting drawings into electronic sound, chapter 3 teases out how each explained human capacities through musical machinery. Together, Diderot and Oram point to musical instruments as a source of alternatives to prevailing paradigms of how humans are and are not like machines.

Moving into the twenty-first century, chapter 4 uses reactions to pianos becoming trash to examine the special status of pianos as more than inanimate objects, closer to a pet or loved one than a piece of furniture or tool. For those who feel this phenomenon, the violent, inhumane end pianos meet stands uniquely equipped not only to shock (as in Annea Lockwood's "Piano Burning" and the MIT piano drop) but also to call into question assumptions about animate versus inanimate matter, about the soul and subjectivity, and about the prerequisites for consideration as a person. The long history of recognizing a person-like status for domestic keyboard instruments helps explain shocked reactions to piano dumping and exemplifies crucial differences between anthropomorphizing and personifying. Shocked reactions additionally reflect relationships to piano materiality that have changed with the piano industry and market and with the specific conditions of the early twenty-first century, when the status of acoustic pianos is tied up with their digital substitutes.

Chapter 5 takes up the sound of chopped and pitched vocals, which "took over pop music" around 2016 and inspired music critics to reassess what sounds human. Tracing a transformation in music critics' perceptions of Auto-Tuned vocals—from the sound of dehumanized robots to posthuman cyborgs—the chapter examines the mainstreaming of posthumanist theories within pop music discourse leading up to and through the ca. 2016 moment. While salutary in its appreciation of artistry in sounds previously maligned as robotic, however, the result has also been a flattening, an assimilation of machinic manipulations of voice into the singer's humanity. Left aside have been the Black posthumanist critique of the category *the human*, which initially attended the flip of dehumanization into a valorized posthumanization, and the division of labor between singer and producer, which disrupts any one-to-one mapping between a chopped and pitched voice and a person. Including such factors helps make audible different aesthetics within the chopped and pitched sound of 2016—in hit songs by Calvin Harris and Rihanna, Felix Snow and Kiiara, and MNEK and Zara Larsson—along with different versions of what it means to be, different genres of being posthuman.

A coda turns to how some musicians are working with machine learning systems in ways that challenge and move beyond the conventional framings of artificial intelligence (AI) as representing either an existential threat to humans or just another tool to serve their creativity and self-expression. Artists such as Dadabots, Holly Herndon, and Arca have not only explored the sonic possibilities of ca. 2020 machine learning but also used their experiences to offer expanded vocabularies for thinking about these computational systems. Their work helps move beyond a narrow fixation on how humanly machines can perform (a fixation built into the very notion of "artificial intelligence") to questions of how machines perform and what it means to work with and exist in relation to them.

"BUT WHAT CAN THIS MEAN?"

Sounding Human thus dwells on musical case studies that help cultivate a wider field of possible relations between human and machine, a project with implications for both imagining the future and interpreting the past. Consider Giovanni Battista Pergolesi's *La serva padrona*, the opera best known for setting off a pamphlet war, at once aesthetic and political, between the supporters of French opera and of Italian opera in 1750s Paris. In this comedy, Serpina, a servant, tricks her master, Uberto, into marrying her. (As the *Mercure de France* described, "Uberto is an old man dominated by Serpina, his maid."[54]) Despite their mutual exasperation throughout the opera, by the end, they both embrace their new life as a married couple, and in the finale, they express their love for one another. Analyzing its reception, Wye Allanbrook has shown how *La serva padrona* struck its philosophical advocates as "a natural organism" manifesting "a new vitalism," in contrast to the "Cartesian machines" of French opera.[55] This was an opera alive with (as Diderot put it in *Rameau's Nephew*) "the animal cry of passion," which in opera "should dictate the melodic line, and its expressions should be pressed out urgently, one after the other."[56] Defenders of French opera like Élie-Catherine Fréron, meanwhile, appealed not to passionate cries but to the mechanics of vibration and the human body, explaining that nerves, like strings, respond to vibrations with which they are tuned in unison, thereby "allow[ing] the soul to experience sensations."[57]

Given the neat opposition between organism and machine, expressive vocal cry and mechanically vibrating nerves, that appears to line up behind Italian and French opera, it is surprising to hear how Serpina and Uberto musically express their love in the opera's finale.[58] Serpina figures herself as the bell of a clock, singing that she has "within my heart / love's little hammer / that strikes me every hour." Uberto replies that he has within his heart "love's big bass drum / that beats strongly every hour." They direct each other to listen, to hear the sounds of their respective clockwork hearts. "Listen: tipiti, tipiti, tipiti," Serpina sings, her vocal imitation of a bell leaping up and down the interval of a fourth (E–A). Her singing is echoed by pizzicato violins, now the direct sound of her heart, followed by a silent pause in which to listen to fading vibrations of the last bell-imitative pluck. Uberto responds, "Listen: tapatà, tapatà, tapatà," his vocal imitation of a bass drum sounding the same interval as Serpina but on the tonic rather than the dominant harmony of the number. His vocal imitation of the bass drum is also echoed by pizzicato violins, reinforced two octaves below by pizzicato bass. Again, a silent pause allows a moment to listen to the reverberance of the final pluck. "But what can this mean?" the two then sing together about their tippitì/tapata-sounding hearts. The answer: "I don't know.... Well you can imagine" (fig. 0.1; audio 5).

What can we imagine this means? In Serpina's leaping bell-like fourths, we might imagine a precursor to later female opera characters—the Queen of the Night from Mozart's *Magic Flute*, the android Olympia from Offenbach's *Tales of Hoffmann*—whose leaping vocals became the sound of singing drained of humanity, all machine (audio 6 and 7).[59] In *La serva padrona*, however, Serpina's leaping voice does not mark her as Other to the opera's human characters. Instead, Uberto sounds melodically the same, while different in register and vowel resonance, and these are the sounds of their feeling responsiveness to one another. Perhaps, then, we can imagine that Pergolesi has offered a musical version of materialist theories in which human emotions and selves can be explained by clockwork mechanisms. Since this is comedy, however, we may also imagine that we should not take this representation seriously—that it is funny precisely because it is so incongruous with how we expect music to express emotions and how we experience ourselves. Contemplating Serpina and Uberto's duet as comedy, we might recall Henri Bergson's theory of the comic as "something mechanical encrusted on the living." Formulated around 1900, this phrase has often been lifted from its context to serve as a kind of universally valid insight into the humor of the mechanical. Its "unquestioned logic," Chude Sokei demonstrates in his reading of Bergson's essay, is rooted in a "racist technopoetics" that conflates the mechanical with Blackness and is incapable of imagining Black subjectivity.[60] One need not venture too far from *La serva padrona* to find this logic on operatic display: a genealogical line could take us, again, to Mozart's *Magic Flute*, this time to its portrayal of the Black character Monostatos and slaves dancing automatically to the sound of bells (audio 8).[61] What Pergolesi presents, however, is something

FIGURE 0.1. Serpina and Uberto sound their love in the final duet, "Per te ho io nel core," of *La serva Padrona*, an intermezzo by Giovanni Battista Pergolesi performed in Paris in 1752 (Paris: Auguste Le Duc, undated). University of North Texas Libraries, UNT Music Library.

mechanical at its main characters' very hearts. Rather than asking us to decide whether they are human or machine or are human despite their mechanical makeup, the duet stages something more like Serpina and Uberto being human by way of machines.

The obvious and the outside possibilities in how human-machine relations can be imagined have changed. The study of history can show us such change; it can also contribute to bringing it about. As a site of real and imaginary relations between humans and machines, *human* and *machine*, music has been central to the experience, definition, and transformation of these categories. But as the examples in this book show, history is not unfolding in a unidirectional sweep toward the ever more mediated or technological to arrive ultimately at either AI takeover or immortal consciousness on a hard drive. Rather, it is unfolding in an uneven but continual process of interpretation and reinterpretation, negotiation and renegotiation. By excavating the making and breaking of human *or* machine logic, we may come to better understand how music can limit as well as expand the possibilities for imagining and enacting new configurations.

ONE

Becoming Android
Reinterpreting the Automaton Flute Player

In September 1739, Françoise de Graffigny took Voltaire and Émilie du Châtelet sightseeing in Paris. Their itinerary included a visit "to the flute player, and it gave me great pleasure," Graffigny recounted in a letter, referring to the flute-playing automaton by Jacques Vaucanson on display at the Hôtel de Longueville.¹ Though mentioned only briefly, Vaucanson's flute player stands out from the other curiosities Graffigny noted for the pleasure it gave her. By contrast, she observed that the manuscripts she saw at the Royal Library "would have delighted a savant but did nothing for me." And about Louis Bertrand Castel's ocular harpsichord (which substituted colors for sounds), she wrote that it was "the dullest and most ridiculous invention that human extravagance has ever produced."²

Pleasure, delight, admiration: such are the sentiments that predominate in the early reception of Vaucanson's flute player. The popular magazine *Mercure de France* reported that "all of Paris has been going to see with admiration a phenomenon of mechanics, the most singular and at the same time the most pleasing that has perhaps ever been seen."³ A decade later, a celebratory poem praised Vaucanson for "putting Nature & Art in unison" and described how, "in the happy assembly of thankless materials, / I hear the sweet concerts of a famous flute player / Who, surprising the mind, charms and delights the heart."⁴ The pleasure surrounding Vaucanson's flute player in its first decade stands in stark contrast to the anxieties later associated with musical androids. The affective difference has led scholars to characterize the eighteenth-century objects as "playthings of the aristocracy" and to interpret the lack of "a generalized *angst*, that ponders them in philosophical terms" as the mark of an audience "largely indifferent to philosophical overtones."⁵

Rather than indifferent, however, we might find mid-eighteenth-century audience members like Graffigny attuned to "philosophical overtones" of a different type—those that were experienced and expressed through pleasure rather than dread. As Georgia Cowart reminds us, "pleasure also motivated more serious philosophies and aesthetic systems."[6] Though at the same time, insisting on a certain seriousness may also be part of the problem; as Robert Darnton has observed, for anyone expecting a "highbrow," "metaphysical view of intellectual life in the eighteenth century," a look at what and how people read in the Age of Enlightenment gives the impression that "a lot of trash somehow got mixed up in the eighteenth-century idea of philosophy."[7] Such an alternative view suggests there is more to the changing significance of musical automata than either an awakening of philosophical reflection or what can be gleaned from their deployment in overtly philosophical texts like those of Julien Offray de La Mettrie and Diderot.[8]

As a technical object, meanwhile, Vaucanson's flute player occupies a pivotal historical position; it has been described as both the culmination of an early modern tradition of automata that "aspired to ever more exact replicas... of living creatures" (by Lorraine Daston and Katherine Park) and as "the first actual android of the new, experimental-philosophical variety" (by Riskin).[9] The latter claim has secured the flute player canonical status in the history of robotics and artificial life, making it the inaugural machine in a tradition of simulating living processes by mechanical means.[10] In rendering Vaucanson's flute player a familiar sort of android, however, this historical perspective has largely missed the cultural dynamics that shaped this automaton and its early reception, which differed from while also setting in motion the later history of human-simulating machines.

Graffigny's writings open onto an alternative history of Vaucanson's flute player, and we shall follow her as a guide to encounters with musical automata—literal, fictional, and figurative. Graffigny was born in 1695 into a noble family in Lorraine, then an independent duchy. An ill-fated marriage left her an impoverished widow at the age of twenty-five, yet with the protection of the ducal court. Per the Treaty of Vienna in 1736, however, Lorraine became a province of France, and its court was disbanded, throwing Graffigny into precarity. In February 1739, she arrived in Paris to become lady-in-waiting to the duchesse de Richelieu. From Paris, she wrote letters nearly daily to François Antoine Devaux, her friend who remained in Lorraine. As English Showalter, Graffigny's biographer and editor of her correspondence, notes, these letters give "an astonishing picture of a woman's life; she told Devaux about almost everything, from her writing projects to her sex life, her finances, her health, her pastimes, her ideas, her feelings, and the news of literary Paris in the heyday of the Enlightenment."[11] In 1747, Graffigny published *Lettres d'une Péruvienne*, a best-selling novel that made her the "most famous living woman writer" until her death in 1758. A sentimental fiction from the fantasized perspective of a Peruvian woman kidnapped by conquistadors, *Lettres d'une Péruvienne* dealt with practices of relating exterior signs and interior life—of questioning humanity—in a colonial frame that was newly converging with Vaucanson's automaton. Although the novel remained well-known into the early nineteenth century, Graffigny was then largely forgotten, appearing mainly in footnotes on Voltaire until feminist literary scholars redirected attention to women writers.[11]

In her study of musical automata from the late eighteenth century, Voskuhl examines how these artisanal objects became literary motifs that functioned "to destabilize the boundary between humans and machines, challenging readers' notions about what distinguished the two."[12] Following Graffigny's lead, however, suggests that in the earlier Enlightenment moment (the late 1730s–1740s) in which Vaucanson's flute player was displayed and she was writing, the idea of a "human-machine boundary" was not so firmly in place as to be a target of destabilization. Rather, the musical automaton functioned in a context of sentimental sympathy in the manner Festa has described for sentimental literature: "Sentimental sympathy does not depend upon a preexisting category—instead, it proceeds on a case-by-case basis, finding and affirming humanity based on the ability to excite or experience emotion."[13] Graffigny's writings reflect her firsthand experiences not only with finding and affirming humanity in others but also with using "the ability to excite or experience emotion" to perform her own humanity in a social world increasingly concerned with such proofs. The intersection of Vaucanson's flute player and Graffigny's letters illuminates a distinctive and formative moment in the history of automata, a phase in the process of their "becoming android" not just in the sense of simulating human life, but in their cultural identity as objects more destabilizing than pleasing to encounter.

A STATUE THAT PLAYS THE FLUTE

When the *Mercure de France* reported on the public display of Vaucanson's flute player, it began its description of the object: "It is the representation in wood of a faun of natural size, elevated on a pedestal that is proportional but nevertheless a little high, in order to contain a whole movement as complicated as the one we are going to explain; the whole exterior is painted the color of white marble"[14] (fig. 1.1). The reason for this white marble appearance became clear a few lines later: "It is an exact and very well rendered copy of the faun executed in marble by the late

FIGURE 1.1. Vaucanson's automaton flute-player, as rendered by Hubert-François Gravelot and François Vivares for the frontispiece to Jacques Vaucanson, *Le Mécanisme du Fluteur Automate, Presenté a Messieurs de L'Académie Royale des Sciences* (Paris: Guerin, 1738). Bibliothèque nationale de France, département Arsenal, 4-S-4798. Source: http://gallica.bnf.fr.

Mr. Coysevox, famous sculptor, and which appears at the end of the great Terrace of the Tuileries" (fig. 1.2). In his written account of the flute player's mechanism, presented to the Académie Royale des Sciences and available for purchase where the flute player was displayed, Vaucanson emphasized how accurately he reproduced the actions—the breathing, fingering, and mouth positions—of "a living man" (*l'homme vivant*).[15] Yet in outward appearance, Vaucanson's flute player did not attempt to match a living human; rather, it was made to look like a marble statue to which capacities of movement were added.

The statue by Antoine Coysevox, *Faun Playing the Flute* (1709), was part of a trio commissioned in 1707 and placed in 1716 in the royal gardens at the Tuileries, where it was accessible to the public.¹⁶ Though the statue lacked the special bodily features that signified a faun (such as pointy ears and horns), its accoutrements—the leafy crown and animal skins—indicated its identity as a denizen of the woods. In court ballets of the period, fauns often appeared in pastoral scenes, where the mythological creatures served to symbolize an "idealized natural world."¹⁷ Throughout the eighteenth

FIGURE 1.2. The statue of a faun playing the flute by Antoine Coysevox, 1709. Photo credit: Archive Timothy McCarthy / Art Resource, NY.

century, Coysevox's white marble flute-playing faun was seen to embody masculine virtues that forest living permitted to flourish. Upon Coysevox's death in 1721, the medical doctor Jean-Baptiste Fermel'huis remarked in a eulogy delivered to the Académie Royale de Peinture et de Sculpture that Coysevox "expressed in the Faun the vigor and strength of a country man" (*a exprimé dans le Faune la vigueur et la force d'un homme champêtre*).[18] In his *Voyage pittoresque des environs de Paris* (1752), Antoine Nicolas Dézallier D'Argenville echoed Fermel'huis's characterization, describing the faun as "symbolizing the strength and vigor of a man from the fields and forests" (*symbolisant la force et la vigueur d'un homme des champs et des forêts*).[19]

Several differences between Coysevox's statue and Vaucanson's automaton copy (as rendered by Hubert-François Gravelot and François Vivares for the frontispiece of Vaucanson's treatise) attenuate the former's look of muscularity and Classical antiquity and give the latter an appearance more fit for the eighteenth-century musical stage (see figs. 1.1 and 1.2).[20] The statue's upper body and feet are bare, whereas the automaton is fully clothed and appears to be wearing boots. The animal skin of the statue is replaced by what appears to be a textile product, adorned with foliage-style decorations. The statue is bearded, while the automaton's face is hairless. In place of oak foliage atop the figure's head, we see a feathered hat, the plumage resembling headwear commonly seen in operatic costumes. The result is thus less a literal copy of Coysevox's statue than its adaptation to theatrical presentation.

Vaucanson did not publish any explanation as to why he chose Coysevox's statue as a model for his flute player. Following a chain of symbolic substitutions, Daniel Cottom suggests that the sculptor's "rivalrous refashionings of antique masterpieces" placed him and Louis XIV's court in relation to Classical antiquity, while Vaucanson "effectively outmarveled Coysevox's pastoral" and "introduced his modern self into the courtly icon."[21] Other aesthetic factors may have contributed to the choice. For one, the statue was poised in the act of flute playing in so lifelike a manner as to make the sound of his playing palpably absent—or else virtually present—either way inviting a realization of musical sound to complete the image. As Fermel'huis remarked in his eulogy, Coysevox "has imitated so well the way in which one can draw pleasant sounds from the instrument he plays, that one can believe that we are missing something to hear it."[22] Dézallier D'Argenville imagined the faun listening to his own playing: "He turns to his left his head whose frizzy hair is tangled with oak foliage; his eyes held on his instrument, he lovingly follows the harmony of sounds modulated by his agile fingers."[23]

The flute also speaks to Vaucanson's design of the automaton for a dual audience: the scientific community represented by the Académie Royale des Sciences and a paying public with various motivations and priorities. The interests of the former are at stake in Vaucanson's explanation of his choice of instrument in terms of technical challenge and discovery: the indeterminate embouchure (and hence the varied angle of the air supply) made playing a transverse flute more challenging to mechanize than playing a recorder, with its determinate embouchure. (The Académie Royale des Sciences recognized Vaucanson's work with a certificate of commendation.) For the public, it mattered that the transverse flute was in vogue. The instrument had joined and gradually begun to supplant the recorder at the Paris Ópera and other courtly music ensembles in the 1670s. The first treatise published in France on playing the transverse flute, Jacques Hotteterre's *Principes de la Flûte Traversière, de la Flûte à Bec, et du Haut-Bois* (*Principles of the flute, recorder & oboe*, 1707), appeared just two years before Coysevox completed his statue. In it, Hotteterre described the transverse flute as "one of the most pleasant and one of the most fashionable instruments."[24] According to a courtly visitor to Paris who attended privately hosted concerts in the 1710s, "the instruments which are the most popular now in Paris are the harpsichord and the transverse or German flute."[25] With the founding of the Concert Spirituel, a public concert series in 1725, the transverse flute had another venue. Introducing the instrument to that stage in 1726 and appearing regularly from 1731 to 1735, Michel Blavet became its most celebrated virtuoso of the period. When Pierre-François Desfontaines described Vaucanson's flute player, he suggested the faun learned from this master, writing "[it] executes with as much force and elegance as accuracy and precision several symphonic melodies, some of which are quite difficult, such as the Nightingale of Blavet, of whom the faun was the student."[26]

The choice to model the automaton on a statue of any kind, meanwhile, points to the conceptual proximity between these two modes of simulating the human form, before androids acquired a distinct cultural identity of their own. When René Descartes set out to explain the body and soul in his *Traité de l'Homme* (ca. 1632), he asked the reader to imagine men whose bodies are "nothing other than a statue or machine of earth, which God has formed expressly to render it as similar as possible to us."[27] Descartes argued that animals thus made "as similar as possible" would be indistinguishable from their originals. Such "statue" or "machine" simulations of men, by contrast, would remain discernable thanks to their lack of two uniquely human capacities that required an immaterial soul: the use of language and of reason.

Vaucanson's audiences would likely have had less experience with

Descartes's thought experiments, however, than with the statue scenario from Ovid's *Metamorphoses*, which supplied a context for witnessing an inanimate human figure acquiring movement and coming to life. Thomas Tolley has called the eighteenth century "the great age of the Pygmalion legend," as the story of the sculptor in love with his beautiful statue was adapted across artistic media (operas, ballets, paintings, engravings), with a particular focus on the "first stirrings of animation in the statue."[28] Noting how tellings of the Pygmalion story have changed with the "beliefs and prejudices" of their time, J. L. Carr contrasts medieval preoccupations with the immorality of Pygmalion's lust and the terror of magical vivification with an eighteenth-century focus on the process of becoming sensing and thinking and on celebrating the power of love.[29]

As a female figure of beauty who does not play a musical instrument, of course, Pygmalion's statue differs from Vaucanson's flute player in significant respects. Nevertheless, through and around this preeminent instance of an animated statue, we may gain insight into the elements of familiarity—familiar symbols as well as processes of reinterpretation—that helped orient audiences in their experiences of Vaucanson's novel machine. Consider the opera-ballet *Le Triomphe des Arts*, created by librettist Antoine Houdar de La Motte, choreographer Guillaume-Louis Pécour, and composer Michel de La Barre and produced at the Paris Opéra in 1700. The fifth and final act stages the Pygmalion legend and exemplifies Pygmalion's status as a laudatory figure in the eighteenth century; his legend concludes a series of episodes centered on Classical figures, each of whom demonstrates the power of an art (Apollo—architecture, Sappho—poetry, Amphion—music, Appelles—painting, Pygmalion—sculpture). On the frontispiece to the published score, a seated transverse flute player plays his instrument. The image refers to the legend of Amphion, the subject of the third act devoted to the power of music. In his preface to the libretto, La Motte explained that he had reconfigured existing Amphion legends: "The poets say that he raised the walls of Thebes to the sound of his lyre, and the mythologists that he gathered together the men who had hitherto been scattered in the forests, and that he united them under the laws of a reasonable society." La Motte "join[ed] these two marvels" and added love for his wife-to-be, Niobe, as their "plausible motive."[30] The frontispiece, however, depicts the lyre discarded at Amphion's feet, as he plays a transverse flute to "raise the walls of Thebes" (fig. 1.3).

The iconographic reinterpretation of Amphion as a transverse flute player is realized sonically in La Barre's score. (La Barre himself was a court flutist who played in the orchestra of the Opéra and in 1702 published the first book of pieces specifically for transverse flute.[31]) Through the in-

FIGURE 1.3. Frontispiece depicting Amphion playing a transverse flute (replacing his traditional lyre, shown discarded at his feet), from the story represented in act 3 of La Barre, *Le Triomphe des Arts* (Paris: Christopher Ballard, 1700). Bibliothèque nationale de France, département Musique, VM2-160. Source: http://gallica.bnf.fr.

teraction of sung text and instrumental parts, flutes become an extension of Amphion's voice and provide a sonic link between the legends of wall building and society building, making music an agent of both. The number during which the desert scene is transformed to the city of Thebes features passages for "flûtes seulles" (flutes alone). The first such passage is heard immediately after Amphion sings, "Dreadful dens, dark retreats / Let my voice dispel your shadows," establishing the flutes as Amphion's voice (fig. 1.4). This metonymy still resounds when, after the last passage for flutes alone, he turns from addressing the desert landscape to addressing "you, wild mortals" (*vous, sauvages Mortels*): "Come, if my voice draws you" (*venez, si ma voix vous attire*), he sings, redirecting the power of the flute from erecting buildings to drawing those dwelling in the mountains, woods, and country to join "under a fortunate empire" (*sous un Empire heureux*) (fig. 1.5).[32]

The transfer of powers traditionally invested in the lyre (Orpheus's as well as Amphion's) to the flute is emblematic of the broader system of

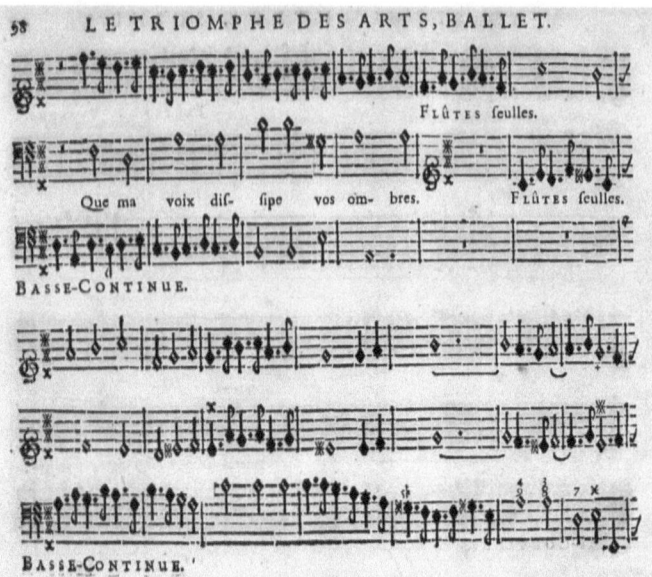

FIGURE 1.4. Amphion sings "Let my voice dispel your shadows," and then flutes sound this voice; La Barre, *Le Triomphe des Arts*, act 3, scene 1 (p. 98).

reversals at work in this opera-ballet. As Cowart has shown, *Le Triomphe des Arts* recast an earlier court ballet that envisioned a subservient role for the liberal arts, their function being to celebrate peace and prosperity achieved through the art of war. *Le Triomphe des Arts* instead portrayed the arts "leading the way to a new, peaceful society under the direct inspiration of Venus."[33] In La Barre's score, flutes provide the sound of Venus's rule. Throughout the work, La Barre associates flutes with Venus's animating power—with the pleasure and love that bring people together and both inspire and are experienced through art. Flutes are specified in act 1, when Venus calls to the pleasures to "show through your play the triumph of the arts" (*faites voir dans vos Jeux le triomphe des Arts*).[34] They also sound when Venus arrives to imbue Pygmalion's statue with life in act 5 (see fig. 1.6), animating the statue (*animer cette image*) and inflaming it with love for Pygmalion (*l'enflamer pour toy*) as a reward for his art (*ainsi que ton Art reçoit sa recompense*).[35]

Within this constellation of figures, Vaucanson's flute player might appear as an amalgam of Amphion (in its musical performance) and Pygmalion's statue (as an art object). The scenarios overlap in the sound of the flute as the sound of pleasure, love, animation; they also share a role for spectators who are praiseworthy insofar as they are responsive to these forces. Prior to Venus's intervention, Pygmalion reflects on his own failings: "I feared to be sensitive, I had to be punished for it. But should I become so for an object that cannot be so?" (*J'ay craint d'être sensible, il falloit m'en punir. Mais devais-je le devenir Pour un objet qui ne peut l'être?*). It is thus not Pygmalion's wayward desire for an inanimate object but his previous lack of feeling that is the site of his moral failure. (These lines were reused verbatim by Ballot de Sovot for Jean-Philippe Rameau's *Pygmalion*, which premiered in 1748.) To take pleasure in Vaucanson's flute player, then, might be to celebrate not just Vaucanson's genius but one's own sensibility, a quality to be prized and rewarded.

"I IMAGINED THAT YOU WERE SENSIBLE..."

About six months after her visit to Vaucanson's flute player, Graffigny encountered another automaton performance. This one was fictional—people pretending to be machines—and occurred within a one-act comedy by

FIGURE 1.5. Flutes continue as the sound of Amphion's voice, addressing "You, wild mortals" to leave the woods and unite "under a fortunate empire." La Barre, *Le Triomphe des Arts*, act 3, scene 1 (pp. 102–103).

FIGURE 1.6. Flutes and violins sound the arrival of Venus to animate Pygmalion's statue; La Barre, *Le Triomphe des Arts*, act 5, scene 2 (p. 198).

Germain-François Poullain de Saint-Foix at the Théâtre François.[36] Entitled *L'Oracle*, the play features statues and machines that test the conditions of possibility for sympathy and love. As in eighteenth-century tellings of the Pygmalion legend, feeling love for an (ostensibly) inanimate object that cannot return it is something to be rewarded. The statues in de Saint-Foix's scenario, however, have more in common with Vaucanson's automata than with Ovidian myth. Graffigny registered the Vaucanson connection in her account of the play by calling the machines by another name not found in de Saint-Foix's text: automats. *L'Oracle*'s status as comic theater, combined with the use of *statue* and *machine* but not *automaton* or *android* in its text, likely account for the play's absence from the scholarly literature on automata.

L'Oracle features three speaking roles: a fairy queen, her son, Alcindor, and the princess Lucinde. Long ago, the fairy queen received a prophecy from an oracle that Alcindor would only find happiness if he could gain the love of "a young princess who believes he is deaf, mute and unfeeling (*insensible*)." As a lack of sensibility, *insensible* carried implications beyond a lack of sensory sensitivity. To be *insensible* suggested an incapacity for emotion or compassion—the kind of coldness and indifference toward others for which La Motte and La Barre's Pygmalion warranted punishment.

In order to fulfill the prophecy and secure her son's happiness, the fairy queen has raised the young princess Lucinde in her palace away from all humans and served only by statues, which she has shaped from marble and "imprinted with all sorts of movements" (*j'imprimois toutes sortes de mouvements*) by means of her magical powers. These statues include "a little dog that barks after" Lucinde and "a monkey that amuses her with its faces and jumps." The princess has thus been persuaded that she and the fairy are the only two beings who speak, think, know, and reason (*qui parlent, qui pensent, qui connoissent, & qui raisonnent*) and that all other beings have been formed only to serve or amuse them and are "absolutely insensible, without knowledge, and incapable equally of love and hatred, pain and pleasure." Alcindor surmises the implications of his mother's plan: when he is introduced to the princess, she will believe that "I am only a doll, a puppet [*une Poupée, une Marionette*] organized above the ordinary sizes."[37]

With this plot device, Saint-Foix not only set the stage for a romantic comedy but also constructed a dramatic scenario akin to the thought exercise from Descartes's *Traité de l'Homme*, mentioned above. Like Descartes, Saint-Foix used *statue* and *machine* interchangeably when referring to the queen's simulated living beings (the fairy queen first describes the figures as statues made of marble; later, she and Lucinde call them machines). Like Descartes, Saint-Foix casts language and reason as being beyond the fairy queen's powers of simulation (as we shall see, it is autonomous speaking that finally crosses the line and breaks the illusion that Alcindor is a machine). But rather than rely on speaking and reason to differentiate humans from their mechanical simulations, Alcindor posits an instinctual feeling that will allow Lucinde to discern the truth: "Reason can be deceived, but never feeling: her heart will receive from nature opinions that she will taste, without understanding them, and that she will follow by instinct, as the bee will pick the scent of the flowers. This intelligence, this chain, this sympathetic force of hearts will act [. . .] yes, Madam, she will love me, and that day I will become the happiest of mortals."[38] Alcindor thus theorizes a discrepancy between rational and emotional intelligence, the latter being embodied and implanted by nature. This emotional intelligence works through connection (chain, sympathetic force) and by a kind of instinctual knowledge that is common to living beings (like bees) rather than uniquely human.

When we meet Lucinde, she is in a profound reverie, preoccupied with the memory of a kiss on her hand (Alcindor had happened upon her while she slept, kissed her hand, and rushed off just as she awoke). This event has wrought a profound transformation in Lucinde; it seems to her that she "breathes a different air. All nature appears to me more pleasant, more animated."[39] She experiences this new sensibility through her reactions to two birds singing to one another: she admires in them a union and tenderness; when they pause, "they soon start singing again, or rather to answer each other with liveliness, with ardor."[40]

For Lucinde to entertain such notions about others imperils the fulfillment of the prophecy, so the fairy queen tries to disabuse Lucinde of the idea that the two birds are sensible creatures with feelings for one another by drawing an analogy to musical instruments. The birds' responsiveness to each other demonstrated by their coordinated singing is the primary issue, and the queen acknowledges that to respond, one must hear. But, she asks Lucinde, "do you also believe that your harpsichord or your bass viol can hear you, answer you, and are sensitive to the soft accents of your voice, when they match so closely the tones that you sound?"[41] Lucinde scoffs: "Pretty comparison! They are machines." This is precisely the observation the fairy queen requires to make her argument—that the birds likewise are "pure machines, but better organized, because nature ever more industrious, ever more knowledgeable, always superior at art, has composed and arranged the springs herself."[42] In other words, the difference is one of degrees in material organization rather than of kind, and all beings of the "pure machine" kind are *insensible*.[43]

Though Lucinde has heard this argument countless times before, she now refuses to believe it. An "interior sentiment" tells her that the two birds refute the claim; if she could catch them, she would caress them and be attentive to all their needs, whereas she has never had such thoughts about her bass viol or harpsichord nor thought to see if her guitar was cold or hot. Realizing that Lucinde needs stronger proof of the pure mechanism underlying apparent sensibility, the fairy queen offers a different musical demonstration. First, she asks Lucinde to examine a set of three marble statues. Clearly, they have no capacity for feeling—but, the fairy declares, she will make them produce the same movements that Lucinde admires in the birds and mistakes for evidence of their capacity to feel (*qu'ils sentient*) and think (*qu'ils pensent*).[44] The fairy then touches the three statues with her wand, and they spring into action. The middle statue performs movements of surprise and admiration followed by the steps of a saraband. The other two statues supply the music for this dance, one playing a violin, the other (like Vaucanson's flute player) a German flute. An illustration of *L'Oracle* by the celebrated engraver Charles-Nicolas Cochin II shows these

FIGURE 1.7. Illustration of *L'Oracle* (1740), a one-act comedy by G.F. Poullain de Saint-Foix, engraved by Charles-Nicolas Cochin II. Bibliothèque nationale de France, département Estampes et photographie, RESERVE EE-15 (1)-FOL. Source: http://gallica.bnf.fr.

instrumentalists in the background, posed as statues on pedestals (see fig. 1.7). After the saraband, a backstage orchestra joins the flute and violin in playing a cheerful and flowing air, during which the dancing statue becomes more animated by degrees and finally dances a tambourin.[45]

In terms of its music and dance, this episode is exemplary of "scenes of statues 'becoming human' through listening and moving, the statues learning to move gracefully in a variety of ways as the orchestra sounded a series of dance movements," as Ellen Lockhart describes a trope of 1730s–1750s French theater.[46] *L'Oracle*, however, repurposes this process, the point being not to witness a transformation from partially to fully human, but rather to be convinced that the machine body remains unchanged in its (lack of) sensibility, even as its performance becomes increasingly "lively." The performance is thus designed to magnify the discrepancy between intellectual understanding and felt experience and to resolve that discrepancy in favor of the former. This proof works on Lucinde, who reacts to the divertissement by lowering her eyes and looking sad. The performance, she laments, has "confounded and destroyed the ideas which I entertained with pleasure... Ah my poor little birds, are you nothing but machines? I imagined that you were sensible, and that you tasted an infinite satisfaction being together."[47] In addition to the instrumentalists' machineness, then, the divertissement stages Lucinde's sensibility, which makes imagining another's sensibility a source of pleasure. Lucinde's remarks also complicate Alcindor's theory of emotional intelligence; to his "corporeal sensibility," which casts feeling as an entirely passive process, she adds "imaginative sensibility," in which she plays an active role through the formation of ideas.[48]

Reaching for reason to hope, Lucinde wonders if there may be "some being of my species with whom I am destined to live, as these birds live together," and if whoever kissed her hand while she slept might be this very being. The fairy admits there has been a young man around the palace. "Are men also machines?" Lucinde asks (*Les Hommes sont-ils aussi des Machines?*). The fairy answers that they are, though they are more complex (*achevées*) than even the monkey with which Lucinde is familiar. Finally, the fairy queen introduces Lucinde to Alcindor, who pretends to be an insensible machine. Lucinde names him Charmant, ties a ribbon around his neck, and plays with him as if he's a dog. "I speak to him, as if he can hear me and respond," she says; "I like this illusion." When Lucinde asks the fairy queen to "animate" Charmant, to make him "able to think, to speak to me, to understand me, and respond to me," the queen replies with Cartesian reasoning that this is impossible, that the springs can be arranged to imitate some of our actions but never to produce a thought. She explains, "he will go, come, laugh, cry, throw himself at your knees, seem tender, submissive, complacent, in love, worried, and this mechanically, like all those of his species."[49]

To further the pleasure of her illusion, Lucinde wishes to hear Charmant sing. This is within the capacities of the machine, though the fairy queen cautions Lucinde to remember that "these parrots have only a jargon, a series of common words that they pronounce at random, and that they repeat to almost all women indifferently, as they have learned them." The fairy queen then cues the machine to sing by supplying the first phrase: "Everything that breathes…" At this, Alcindor "seems shaken, moved, and like a man who wakes up." He sings:

> Everything that breathes
> recognizes the power
> of charming love.
>
> (*Tout ce qui respire / Reconnoît l'empire / Du charmant amour*).[50]

The source of these lines was a choral number from the opera *Pirame et Thisbé* by librettist Jean-Louis-Ignace de La Serre and composers François Francoeur and François Rebel. *Pirame et Thisbé* premiered at the Académie Royale de Musique on October 17, 1726, but its reprisal on January 26, 1740, made it a timely source for musical quotation in *L'Oracle*. In the opera, the choral number occurs in the prologue and accompanies the descent of Venus into the Temple of Glory. It forms part of the "harmonious sounds" that "fill the air," instrumentally realized with flutes (continuing the sonic association discussed in *Le Triomphe des Arts*) (see fig. 1.8). Glory initially reacts to these sounds as an intrusion, the "pleasures" of Venus not belonging in her temple. Venus, however, reminds Glory that "tender love" can "increase courage" and introduces the story of Pyramus as one of being both lover and warrior. Through this story, the opera develops the idea of tender love and military glory as twin forces of empire, with both lands and hearts as their territories.

The Chorus of Graces sings the phrase "tout ce qui respire" four times to the same melody, making it a musically memorable choice for quotation, while the context of *L'Oracle* charges the line with new meaning (audio 9). The claim that "everything that breathes recognizes the power of charming love" supports Lucinde's intuition about the pair of birds and why she feels differently about them than about her harpsichord and bass viol. Would it extend to Vaucanson's flute player with its mechanized breath, its bellows simulating human lungs? Might it, despite the fairy queen's cautions, encompass the "machine" now singing (whose Lucinde-given name *Charmant* comically transforms the line "empire du charmant amour")? Such lines of philosophical reasoning are cut off by Lucinde's felt response; she exclaims, "The sound of his voice penetrates straight to my heart!" For Lucinde in this moment, the words seem not to matter, the feeling to be instinctual.

Finally, the fairy suggests that Lucinda have fun teaching Charmante

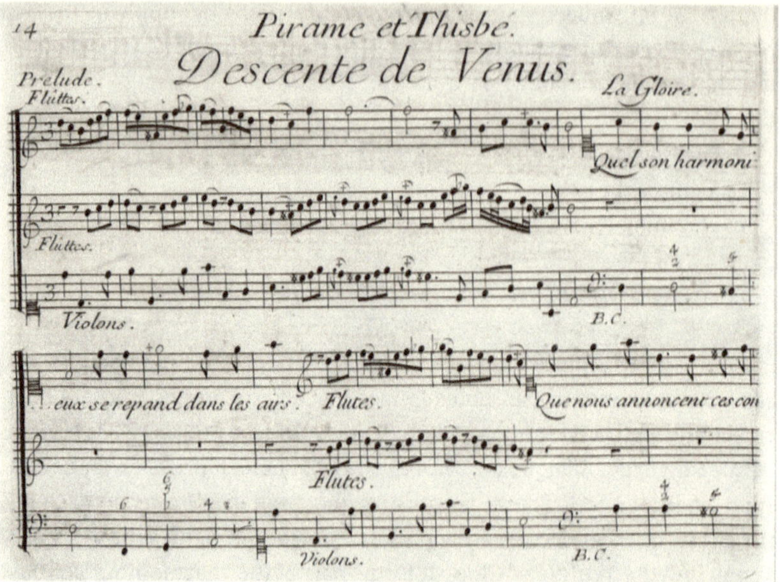
14
Pirame et Thisbé.
Descente de Venus.
Prelude.
Fluttes.
La Gloire.
Quel son harmoni
Fluttes.
Violons.
B.C.
...eux se repand dans les airs.
Flutes.
Que nous annoncent ces con
Flutes.
Violons.
B.C.

Prologue.
15
...certs.
Flutes.
Flutes.
Violons.
Choeur de Graces.
Toutce qui respire, Reconnoit l'empire du
Flutes.

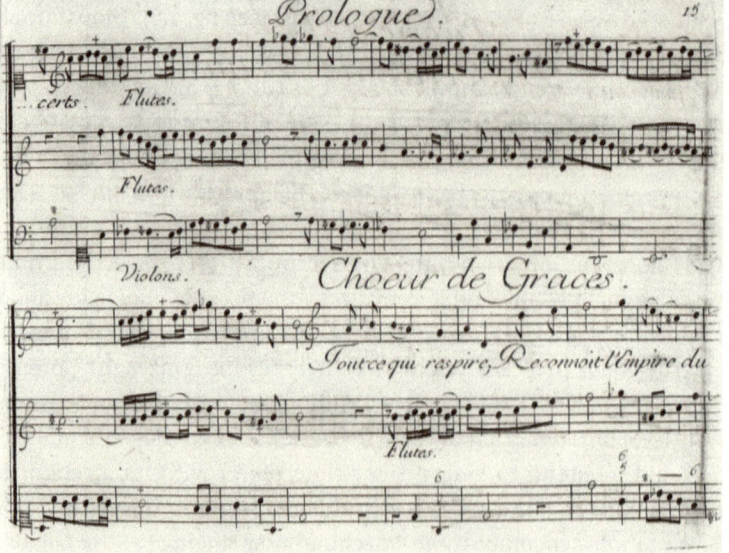

FIGURE 1.8. Venus descends to the sound of flutes, and a chorus of graces sings the lines that "automaton" Alcindor quotes in *L'Oracle*: "Everything that breathes / recognizes the power / of charming love"; from the prologue of *Pirame et Thisbé*, a tragedy set to music by François Francoeur and François Rebel (Paris, 1726). Bibliothèque nationale de France, département Musique, VM2-287. Source: gallica.bnf.fr.

words that he will repeat back to her. The lesson begins with Alcindor repeating after Lucinde: "Lucinde!" "My dear Lucinde!" "I love you." At this third phrase, Alcindor can no longer contain himself and expresses his love in his own words. Lucinde exclaims, "He speaks on his own!" Alcindor finally drops the pretense and declares that the prophecy has been satisfied. Alcindor and Lucinde are happily united, and the fairy queen has the last word: "After having been a deaf, mute and unfeeling lover, be an eager [*empressé*], tender [*tendre*], and obliging [*complaisant*] husband."

Graffigny described *L'Oracle* in a letter dated March 27, 1740, to her friend François Antoine Devaux. Her account of the plot is quite detailed yet includes differences from the printed play that are suggestive of her own interpretive lenses (her use of the term *automat*, for instance) or, possibly, of changes made by the performers.[51] For instance, Graffigny omitted the scene of Alcindor singing, which, given her detailed accounting of the play, leaves one to wonder whether it was omitted from the performance. For Graffigny, the highlight of the play was what followed when the fairy let Lucinda play with the prince automaton as she would with her monkey and parakeet. Graffigny describes a scene that displays the emotions of the prince and princess while Lucinda treats the prince as if he's a dog, directing him to come to her, sit, and so on and putting a leash on him. It was this scene that was selected for Cochin's illustration of *L'Oracle* (fig. 1.7). Graffigny details the gestures Lucinde used to communicate directions to her machine-man, which he followed. "All this is accompanied by naïve words that express passion, but the game is so charming that you can't hear the words," Graffigny wrote, suggesting that—as for Lucinde when Charmant sings—it was the voice, not the words, that mattered.

By Graffigny's account, the fairy then tells Lucinda she can teach the automaton to speak, just as she had done with her parakeet. When Alcindor breaks off from repeating Lucinda, she exclaims, "He's talking on his own!" "That's the denouement," Graffigny concludes. "They love each other, the fairy marries them, and they are comfortable." Graffigny's synopsis and enjoyment of Saint-Foix's play confirms what its characters perform: it is pleasurable to imagine other beings to be thinking and feeling, and one's felt intuitions are to be believed above arguments that would deny such sensibility to others. Through its automaton encounters, *L'Oracle* exemplifies what Festa finds in sentimental texts: they "labor to suture the gap between intellectual and emotional recognition" and perform "the cultural and literary labor that aligns the felt perception of who is a person with emerging historical categories that delimit the human."[52] In doing this work, *L'Oracle* (like Vaucanson in fashioning his flute player) figures automata as moving statues. The implications of automata for "felt perception" and "emerging historical categories that delimit the human" would change, however, as moving statues receded in cultural memory and other figures came to the fore.

VAUCANSON'S FLUTE PLAYER BECOMES *SAUVAGE*

Vaucanson's flute player and two automata companions left Paris to tour Europe around 1740, with a sojourn to London from late 1742 to mid-1743.[53] In 1743, Vaucanson sold the set of automata to a trio of businessmen from Lyon who continued its European tour.[54] When the machines eventually returned to France, the flute player had accrued a new identity: *sauvage*. Absent from descriptions of Vaucanson's flute player in the 1730s, the loaded term *sauvage* appears on a publicity poster of unspecified date and locale; however, André Doyon places it securely in Strasbourg in late 1746 (see fig. 1.9).[55] The poster describes the flute player as "a man of natural size in WILD (SAUVAGE) dress who plays Eleven Airs on the Flute by the same movements of keys by fingers & blowing by mouth as a living man." While *sauvage* is an adjective describing the figure's dress, the typeface lifts the word out of its grammatical context to let it simultaneously function as a noun along with the other two automaton nominations: *berger provencal* (Provençal shepherd) and *canard* (duck).

With the descriptor *sauvage*, the poster affiliated the flute player with (to begin with a definition from the official French dictionary of 1694) "certain peoples who ordinarily live in the forests, without religion, without laws, without fixed habitations, and more as beasts than as men."[56] We have seen *sauvage* used with this meaning in *Le Triomphe des Arts* in 1700 to describe the people whom Amphion called from the forests with his flute, so as to bring them into Thebes and a "fortunate empire." By the 1740s, however, *sauvage* more often appeared in opera, as well as the travel and philosophical literature of this colonial era, as the term for people native to the Americas. A year after the premiere of the opera-ballet *Les Indes galantes* in 1735, for instance, Rameau added to it a new entrée entitled "Les Sauvages"; while the other three entrées made a global tour to a "Generous Turk," "The Incas of Peru," and a "Persian Festival," this one was set in "a grove in the forests of America" (i.e., French colonial Louisiana).[57] As Richard Nash has argued, Enlightenment-era discourses on the *sauvage* reflect not simply an error of taking metaphors such as "wild man" and "noble savage" literally, but rather a cultural process in which "a preexisting mythological terminology actually shaped the preconceptions and hence perceptions by which real beings were observed and recognized by Europeans."[58] Yet another facet of this cultural process is reflected in the

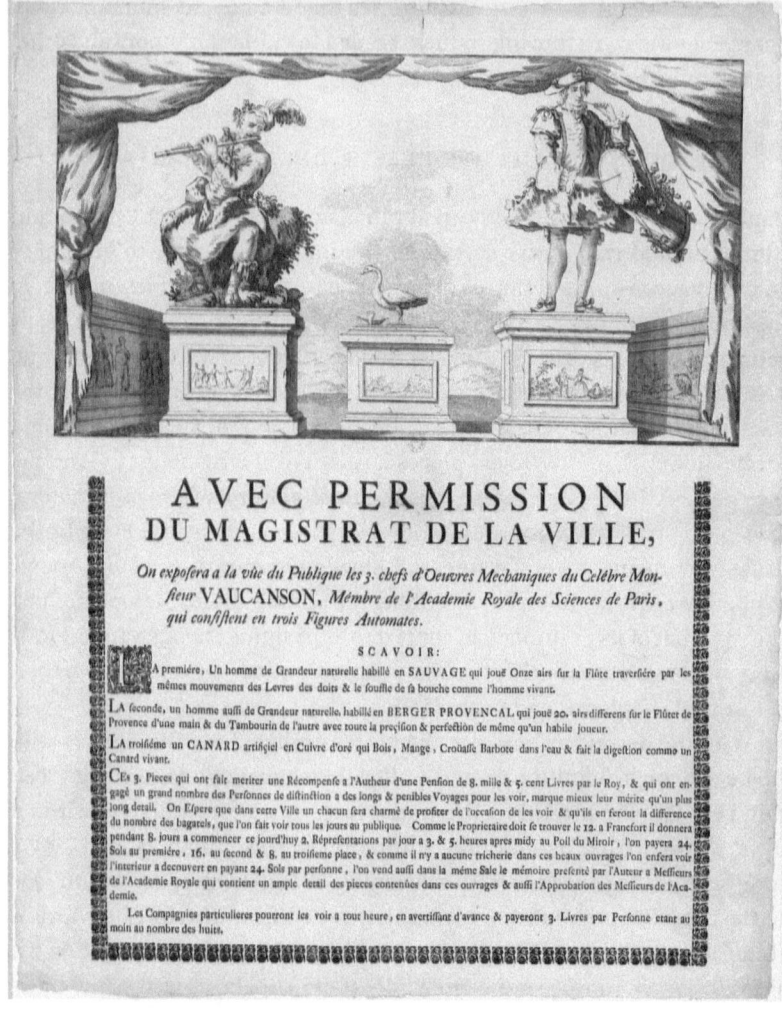

FIGURE 1.9. Poster advertising an exhibition of Vaucanson's automata, 1746. Bibliothèque nationale de France, département Estampes et photographie, RESERVE QB-201 (100)-FOL. Source: http://gallica.bnf.fr.

doubleness of *sauvage*—at once adjective for *dress* and noun for *person*—on the 1746 poster. Scholars have traced a transformation in conceptions of difference in French colonial sources from the late seventeenth to mid-eighteenth centuries, showing—as Sophie White writes—a "transition from a conception of identity as fluid and mutable to an essentialist conception based on proto-biological assumptions."[59] Within the larger geopolitical contexts for this transformation, White demonstrates the centrality of dress to expressions and perceptions of identity in the French American colonies, with attire operating as a site of mutability between *sauvage* and French identities. On the poster, the mutability of *sauvage* dress vies with the growing fixity of the *sauvage* as a racialized Other.

It may be that in bringing the flute player to audiences outside Paris, the model of Coysevox's faun was left behind and the association between automaton and animated statue attenuated. Likely, too, the duck and pipe-and-drum player (which Vaucanson added in 1738, a year after first displaying the flute player) encouraged assuming a model for the flute player in real life rather than art. Into the interpretive space thus created, the *sauvage* was increasingly available to project. In fact, a related transformation took place on the French operatic stage: as Hélène LeClerc has shown, inventories for productions of Rameau's *Indes Galantes* list costumes for fauns to be used as costumes for *sauvages* (the opera's Native American characters).[60] Dance historian Joellen Meglin suggests that "this simple act of substitution of one costume for another, as if they were interchangeable parts, seems to encapsulate a complex intercultural process, where a substitution is a way of interpreting new, relatively unfamiliar theatrical ideas by way of older, more familiar ones." In the case of Rameau's *Indes galantes*, casting the *sauvage américain* as a "modern-day faun" recycled the existing figure of a forest denizen who was "uncultivated but pure, lacking the sins and satiety of civilization."[61] On the 1746 poster, a similar substitution is on display—though I am suggesting it was the older idea of Coysevox's statue that had become unfamiliar, making way for the *sauvage* of the colonial imagination (itself shaped by preexisting mythological figures) to take its place.[62] The poster's description of the flute player as a "man of natural size" modifies the earlier description in the *Mercure de France* of the flute player as a "faun of natural size," overwriting the flute player's former mythological identity with the new one of an *homme sauvage*.

The illustration on the ca. 1746 poster resembles the one by Hubert-François Gravelot that featured as the frontispiece to Vaucanson's explanatory treatise on the automata (fig. 1.10). While apparently modeled on this earlier illustration, however, the image on the poster has also been altered; for one, the portrait orientation of the frontispiece has been adjusted to the landscape orientation of the poster. Beyond such practical adjustments, however, there are several differences between the two illustrations of the flute player that are significant in relation to its new *sauvage* identity. Boots have become sandals; the foliage motif has disappeared

FIGURE 1.10. Frontispiece from Vaucanson, *Le Mécanisme du Fluteur Automate, Presenté a Messieurs de L'Académie Royale des Sciences* (Paris: Guerin, 1738). Bibliothèque nationale de France, département Arsenal, 4-S-4798. Source: http://gallica.bnf.fr.

from the tunic, and there is now more prominent draping of animal pelt around the waist; the face is less smooth in appearance, perhaps bearded; and the plumage droops rather than stands up from the head, conveying less theatrical flair.

Most intriguing, however, is the changed curvature of the left hand. The new curvature is unlike the hand of professional flute players of the period, which—like the left hand of the flute player on the earlier frontispiece—share a more angled wrist and flatter finger position (see fig. 1.11). Discussing hand position in his *Principes de la Flûte Traversière, de la Flûte à Bec, et du Haut-Bois* (1707), Hotteterre noted that "there are some people who place the top [hand] in a different way from how I have shown it, i.e. who hold the wrist outwards (making an arc), and support the flute on the end of the thumb. This hand position does not stop you from playing well, but it is not as natural nor as gracious as the other one."[63] When Antoine Mahaut published a new method (*Nouvelle Méthode*) for learning to play the transverse flute in 1759, he noted there was nothing to add to Hotteterre's explanations of the flutist's body posture and hand positions, which had been "more or less copied" by those who came after him.[64] However, where Hotteterre observed "it is not necessary to observe exactly the rules which I have prescribed for the embouchure and for the position of the hands... you must always follow what seems most natural," Mahaut averred that it is "necessary to prefer the [hand position] we have just taught, being generally adopted by all those who have distinguished themselves on this instrument."[65] From an instrumental newcomer admitting a degree of idiosyncrasy and variability in Hotteterre's day, the transverse flute had become an established instrument with extensive repertoire and codified technique.

The curved hand position thus stands out and calls for interpretation; it would appear wrong compared to "distinguished" flute players, a failure in terms of what is most "gracious," but perhaps also to be "following what seems most natural." Like the *sauvage* label, this hand in the act of music making points away from Coysevox's statue and Vaucanson's work toward the colonial context for experiencing the universality and diversity of the human form, the fluidity or fixity of identity. In doing so, it joins Vaucanson's flute player to the ambivalences of the *sauvage* in the French Enlightenment—its function as "a 'primitive' colonial other at the center of French concerns about the nature of human being and expression."[66] For automaton encounters, the *sauvage* invoked different scenarios and stakes than did an animated statue—dynamics explored by Graffigny in the novel that made her a famous author.

FIGURE 1.11. *From left*: (A) Illustration of flute-playing position from Jacques Hotteterre's *Principes de la Flûte Traversière, de la Flûte à Bec, et du Haut-Bois* (Paris: Ballard, 1722 [first published 1707]), Bibliothèque nationale de France, département Musique, VM8 G-113, source: http://gallica.bnf.fr. (B) Illustration of flute-playing position from Antoine Mahaut, *Nouvelle Méthode pour apprendre en peu de temps a jouer de la flute traversière* (1759), courtesy Russian State Library, https://creativecommons.org/licenses/by-sa/4.0/legalcode. (C) Portrait of Michel Blavet, the celebrated flutist named as the teacher to Vaucanson's automaton flute player, painted by Henri Millot (1720), source: Collection privée / Wikimedia Commons, https://creativecommons.org/licenses/by-sa/3.0/legalcode.

GRAFFIGNY'S ZILIA

While Vaucanson's flute player was being presented as *sauvage*, Graffigny was developing what would become a best-selling novel: *Lettres d'une Péruvienne*.[67] Zilia, the Peruvian woman of the book's title, is betrothed to the Incan ruler Aza. Before their marriage can take place, Zilia is taken captive by Spanish conquistadors, then taken from the Spanish by a Frenchman named Deterville, who brings her to France with his sights set on love and marriage. Like Charles Louis de Montesquieu's *Lettres persanes* (1721, which Graffigny had read), *Lettres d'une Péruvienne* ventriloquizes a non-European Other to comment on French culture and society.[68] As Catherine Liu observes, "Graffigny's fiction of the virtuous outsider relies on the Enlightenment trope of defamiliarization," assigning Zilia to a nation of "nature" and "truth" from which to regard Paris as "the site of falseness, deception, and inauthenticity."[69] We may also read *Lettres d'une Péruvienne* through the lens of Graffigny's encounters with the flute player and *L'Oracle* as a work that finds pleasure in imagining the sensibility of another—specifically, another one has been told does not possess it. As she explained in a preface to Zilia's story, Europeans had failed to recognize the humanity of people from the Americas: "We despise those from the Indies, we are scarcely willing to credit these wretched peoples with a thinking soul; and yet their history is available to us all, and we find recorded on every page of it the astuteness of their minds and the soundness of their philosophy."[70] Like Lucinde in *L'Oracle*, Graffigny finds her experience at odds with what she's been taught; in this case, the source of contradictory experience is a history book.

Graffigny's *Lettres* were written and published during a period of heightened interest in Incan history within France, an interest her personal correspondence partially maps out. She apparently missed Rameau's *Les Indes galantes*, but she saw Voltaire's popular tragedy *Alzire, ou les Américans*, in which the eponymous character is the daughter of a conquered Incan chief slated to marry the son of the Spanish governor, on May 5, 1743 (it premiered in 1736). Like Voltaire, Graffigny read a history of the Incas by Garcilaso de la Vega, himself the son of a Spanish conquistador and an Incan princess. First published in 1609 as *Commentarios reales*, Garcilaso's history was translated and republished throughout the seventeenth and eighteenth centuries. Graffigny likely read a richly illustrated 1737 Amsterdam edition, *Histoire des Yncas, rois du Perou*, which became her main source of information on Peruvian society and culture.[71]

Historical fiction may thus be another genre tag to be added to Graffigny's *Lettres*; the letters ostensibly date from the sixteenth-century Spanish conquest, though the France in which Zilia arrives seems more 1740s. While apparently naïve to the active destruction of native histories by Europeans, Graffigny's novel makes a theme of the material mediations involved in communicating and perceiving a mind across cultural difference.[72] Addressed to her beloved Aza, Zilia's letters (according to Graffigny's historical-fictionalizing frame) were initially written in her native language and quipu, the knotted-string system of Inca writing; only because she later learned French and alphabetic writing was Zilia able to translate her letters for French readers. In the letters, Zilia's transition to French is reflected in her use of the term *sauvage*. Only in her quipu-written letters does Zilia refer to the Spanish and French as *sauvages*. The term does not simply mark European barbarity toward her and her people. (Upon arriving in France, for instance, she comments on how humane "the savages of this country seem."[73]) Rather, decoupled from its associations with living closer to nature, *sauvage* becomes a relational term—a marker of unfamiliarity, of evident difference in customs and behaviors, and of the lack of communicational means by which to learn what explains those differences.

Zilia's letters dwell on the problem of sharing hearts and minds without shared language and across cultural difference. In one, she narrates the experience of being put on display and treated as an object of curiosity; she recounts people laughing at her, touching her—acting as if she were a mindless and insensible object. Another episode recalls the scene in *L'Oracle* where Lucinde instructs the supposedly mindless and insensible Charmant to repeat after her. Zilia describes how Deterville "begins by having me pronounce distinctly some words in his language. As soon as I have repeated after him, *yes, I love you*, or *I promise to be yours*, delight spreads over his face, he kisses my hands with great emotion and with a look of joy which is the very opposite of the gravity that accompanies divine worship."[74] Only after Zilia learns to speak French is she able to communicate to Deterville her feelings of friendship for him, thereby dashing the illusion Deterville sustained of her romantic love for him. For her part, Zilia is surprised and confounded to learn of Deterville's love; how could he love her, she asks, when she is only now able to speak her thoughts to him? Zilia thus exposes the one-sided and fictitious nature of what Deterville imagined to be a "sympathetic force of hearts" (to recall the sentimental logic of *L'Oracle*). The relationship between Deterville and Zilia might also be read as a feminist revision of Pygmalion, in which language replaces love as the agent of transformation and enables the beloved object to show that she has a mind and desires of her own.

Zilia's trials in a new land provide reason for her to seek out "a universal language of sympathetic hearts."[75] Two main resources become important in this regard: body language and music. In the episode when she is put on display, Zilia aims to "persuade them by my bearing that my soul was not quite as different from theirs as my clothes were from their finery."[76] An evening at the opera occasions Zilia to reflect on musical sound: "It must be the case, my dearest Aza, that the intelligence of sounds is universal [*que l'intelligence des sons soit universelle*], because it was no more difficult for me to be moved by the representation of these different passions than if they had been expressed in our own language, and that seems quite natural to me."[77] These ideas of an "intelligence of sounds" and a "language of sympathetic hearts" recall Alcindor's claims in *L'Oracle* about "this intelligence, this chain, this sympathetic force of hearts" that will prevent Lucinde from being deceived.[78] Zilia concludes, "In short, my dearest Aza, everything in this spectacle conforms to nature and humanity."[79]

The problem with theories of music conformed to nature and humanity, of course, is what happens when real-life musical encounters fail to conform (as with the questions raised by the flute player's hand). It is in this context that the sources for Graffigny's fiction become particularly significant: Vaucanson's flute player and *L'Oracle* were likely among her most direct experiences with musical performance by beings whose thinking and feeling were explicitly put in doubt. Like Lucinde's, Graffigny's ability to imagine sensibility in such beings—rather than threatening the security of her own humanity—was sentimental proof of it. Read from this vantage point, *Lettres d'une Péruvienne* joins the flute player in *sauvage* dress as strands in a process of weaving automaton and Other together, of becoming android.

UNCERTAIN HUMANITY

On Graffigny's musical experiences, we also have her account of attending a rehearsal of Rameau's *Dardanus* on November 9, 1739. "I was enchanted by it and immediately discovered myself to be a Ramoneuse," she wrote, though she also found herself so "close to the orchestra" that it hindered hearing the words. Thanks to a foible of the acoustics of the performance space, then, Graffigny's experience of the opera was—like her character Zilia's—of music separated from language.[80] Despite a chaotic rehearsal scene, she concluded, "It all added up to a very peculiar event, and what with all these antics, with the beauty of the music and the magic of Jéliote's voice, I wept bitterly half the time behind my fan."[81] Her weeping would be a surprising response, given the foregoing description, if not for the great emphasis on tears as a sign of sensibility in her cultural milieu.[82]

Festa has argued that it is as much the humanity of the observer as of the observed that is "staged and interrogated" in sentimental encounters.[83] Graffigny's writings suggest she was very aware of her own humanity being "staged and interrogated" in her encounters with other people as well as with things, like Vaucanson's flute player.[84] In such a sentimental context, pleasure, like tears, was freighted with significance. To take pleasure in a musical machine—even to indulge the illusion that it is thinking and feeling—was to affirm one's own sensibility and humanity. Such an arrangement, however, also brought along the "problems [that] unfettered sympathy poses for the Enlightenment subject when shifted to the domain of empire."[85] Refiguring the faun statue as an *homme sauvage* shifted Vaucanson's flute player to the domain of empire, setting the stage for the musical android to become a destabilizing figure that punishes rather than rewards the extension of sympathy.[86]

Androids are often tied to questions about how much of *the human* can be simulated mechanically, which in the eighteenth century leads to debates over dualist versus materialist philosophies. The automaton in these discourses appears much like those in the fairy palace of *L'Oracle*—isolated from a wider social world, objects to be encountered from an initial state of security in one's own humanity. Carolyn Abbate aptly describes the resultant view: "The perfected mechanical man robs us of a prize, our soul, and in so doing injures human individuality and consciousness."[87] According to this way of thinking about musical machines, it is only as a party to materialist theories that deny the soul that the android can be a source of pleasure. Graffigny's letters document another history, in which the flute player entered a world where the humanity of others was routinely in question and one's own humanity in need of performance. To remember Graffigny as a leading figure of the Enlightenment is to recover how differently the sounds of a mechanical flute player could resonate with such conditions, with philosophical overtones experienced through pleasure before anxiety.

TWO

Hybrids
Voice & Resonance

Father Kircher, that famous Jesuit born, it seems, to steal from nature all her secrets, a century ago desired that an instrument would be possible which was both "string" and "wind." He never doubted that a creative artist who cast before the world such a phenomenon would cast new pleasures. This instrument has been fully invented without anyone's being aware of it; it had existed, but no one had noticed it. It had been left up to M. Ferrein to discover it in the organs of the voice, and to prove its existence by a dissertation equally sound and ingenious.[1]

When I encountered this passage in *L'Art du Chant*—a pedagogical treatise by the Paris-based singer and voice teacher Jean-Baptiste Bérard published in 1755—it caught me by surprise. I had heard of Ferrein as the anatomist who introduced *vocal cords* into our vocabulary. An example of a metaphor that lost its metaphorical quality through regular usage, *vocal cords* had been a novel conceptualization of vocal anatomy grounded in Ferrein's argument—so I thought—that the voice is not a kind of wind instrument but rather a kind of string instrument. I had not known Ferrein to be the discoverer of a string-wind instrument in the organs of the voice. Was the voice, for Ferrein, not just a string instrument but a hybrid instrument?

The voice was indeed not just a string instrument, according to Ferrein. But let me admit at the outset that this passage drew my attention not only because it contradicted what I thought I knew about the history of *vocal cords*, but also because I readily saw Ferrein's string-wind instrument as a hybrid—and hybrids are supercharged with significance for the ontology, historiography, and future of *the human*. In modern times, the voice has been a potent site for experiencing humanity as either pure or compromised by nonhuman elements. Consider Abbate's reading of Papageno's music in Mozart's opera *The Magic Flute* (1791). The recipient of a set of magic bells, Papageno sings an aria, "Ein Mädchen oder Weibchen," in which his singing and bells alternate (audio 10). In this pattern of alternation, Abbate detects a late eighteenth-century version of a cyborg voice, a "chimerical sound-object" that "combines mechanistic and natural human sound." The perception of bell-voice alternation as machine-man hybridization sets the stage for Abbate's hearing of Papageno's later aria, "Klinget Glöckchen, klinget," when Papageno at last sings simultaneously with his bells (audio 11). In Abbate's interpretation, this simultaneity sounds "a miraculous escape from all machines," "the threat of metamorphosis" being "laid to rest in the instant that the bells are reduced to accompaniment, something that has no power to replace or arrest human singing."[2]

Ferrein's theorization of the voice in 1740 is closer to the "pleasure in the confusion of boundaries" that Haraway called for in her cyborg manifesto than to Abbate's "threat of metamorphosis" or replacement. To get at why and with what consequences, this chapter examines "purifying" and "hybridizing" practices around the voice in the late seventeenth to early eighteenth centuries, tracing them through encounters with two particular kinds of instruments: viols and bells. "Pretty comparison, they are machines!," we may recall Lucinde replied to her fairy mother's proposal that birds singing responsively to one another was no different from the responsiveness of Lucinde's harpsichord or viola da gamba to her singing voice. A long Aristotelian tradition conceptualized the voice as the animate instrument, the sound of a being with soul, making animate voice versus inanimate instrument an available configuration throughout the early modern era. Writings around 1740, however, show a significant reconfiguration in practices of listening and thinking about materiality, creating a kind of passageway between earlier conceptions of the voice as the instrument fashioned by God the artificer and later conceptions of the voice as the sound of a pure, natural humanity. To put it another way, sounds, instruments, theories, and listening practices around 1740 formed the conditions from which thinkers like Jean-Jacques Rousseau extracted the voice as *purely human.*

Crucially, the period around 1740 was one of enthusiasm for hybrids—of intellectual and embodied attraction to hybrid beings. Both Ferrein's argument for the voice as a string-wind instrument and the historical forgetting of that argument are exemplary of patterns in the production and disappearance of hybrids. But there is a being from this period that bespeaks the appeal of hybrids and has not been forgotten: Trembley's polyps. An aquatic creature identified by Abraham Trembley in 1739, polyps defied categorization as either animal or plant. Though giving an "initial impression" of being a plant, thanks to their green color, shape, and apparent immobility, sustained observation revealed the polyps to have self-locomotive capabilities, which (in Trembley's own account of his scientific observations) "rouse sharply in my mind the image of an animal."[3] Persuaded that they were animals, Trembley nonetheless carried out a further test of taxonomic identity by cutting polyps into two parts. To his surprise, both parts continued to live and regenerate, like a plant. And yet still, he recorded, "the more I observed whole polyps, and even the two parts in which the reproduction just described took place, the more their activity called to mind the image of an animal."[4] Trembley reflected that "in spite of my impatience to know precisely in which class the polyps belonged, I nevertheless experienced some pleasure in this doubt," which piqued his "curiosity more and more" and led him to make remarkable discoveries.[5]

The implications of Trembley's discoveries—of a creature that could move by its own agency (like an animal) but could also be segmented with both parts continuing to live (like a plant)—spurred substantial scientific and philosophical debates, with lasting consequences for the development of biology that have made Trembley a well-studied figure in the history of science.[6] Trembley's polyps have also been noticed by musicologists, who have found connections to changing eighteenth-century music aesthetics. For Daniel Chua, the polyp instigated a newly organic conception of life and of music—a new biology of the soul. In this view, both living beings and music are characterized by a "formative impulse" that comes from within and unifies its dynamic, plural, proliferating expressions. While Chua finds the realization of this organic quality in late eighteenth-century German instrumental music, Allanbrook locates it in Italian opera buffa. Finding an explicit comparison between polyps and the style of opera buffa in Diderot's *Le Neveu de Rameau*, Allanbrook argues that it was a polyp-like nature that *philosophes* had in mind when they described the style of Italian opera as natural: "Unlike those Cartesian machines of the *tragédie lyrique* that projected the motions of the passions," Italian comic opera "manifested a new vitalism ... the volatility of its matter."[7] The polyp's implications for the organization of matter, Chua and Allanbrook argue, extended to the organization of music.

It is by turning to musical instruments and perceptions of sonic quality, however, that we find something more akin to Trembley's pleasure in taxonomic undecidability, in the mingling of ostensibly distinct categories. Against a common view of the Enlightenment as insistent on sorting things into rigidly grid-like taxonomic identities, Allanbrook notes that some eighteenth-century thinkers were in fact "quite flexible in their allowance for categorical mixture."[8] From Le Blanc's defense of the viol and Ferrein's theory of the voice discussed in this chapter, we see something more: beyond allowing mixture to take place, Le Blanc and Ferrein actively constructed taxonomic categories, newly defining them *in order* to make their object of interest (viol and voice, respectively) a product of mixture.

That these could be mixtures of organism and machine, human and nonhuman, was not self-evident. While Papageno's voice-bell alternation was, in Abbate's reading, as near a cyborg voice as could be achieved with "the technological resources of 1791," a sound that combines machine and organism is not only a question of technological resources; it is also a question of conceptual resources.[9] As we shall see, voice and bell did not sound "entirely distinct ontological zones" (were not "opposed in a metaphysical binary," in Jonathan De Souza's apt phrase for voice-instrument relations) in the first half of the eighteenth century.[10] But the makings of such a binary were coalescing around 1740, as can be discerned in Le Blanc's defense of the viol against the rising popularity of the violin. For Le Blanc, the sound of the viol (played with a modern technique) and certain bells achieved an ideal balance of voice and resonance. Le Blanc explained this mixture of sound quality using terms for breed and race, mixtures of intense scrutiny in his milieu among natural philosophers seeking to explain processes of generation and officials making colonial policies that regulated relations among the free and enslaved. Attending to the technological and conceptual resources at play in listening to instrumental sound and theorizing sound production reveals how hybrids have shaped the history of sounding human.

VOICE AND LUTE

Throughout the seventeenth and eighteenth centuries, when a writer wanted to make the case for their favored musical instrument, they typically deployed one of two arguments—that the instrument was capable of the greatest number and diversity of sounds (see champions of the organ and harpsichord) or that the instrument came closest to the ideal of the human voice. For the viol, being the most voice-like was a go-to claim. As the Jesuit-educated Marin Mersenne observed in his *Harmonie universelle* (1636), "It is certain that if the instruments are taken in proportion that they best imitate the voice, and if of all the artifices one esteems most that which best represents the natural, it seems that one must not refuse the prize to the viol, which imitates the voice in all its modulations, and even in its accents most significant of sadness and joy."[11] Mersenne's comments on nature and artifice reflect a Neoplatonic-Christian understanding of these categories: the voice was the "natural" instrument because it had been fashioned by God the creator; it thus provided the model for all those "artificial" musical instruments fashioned by human hands. Mersenne's contemporary Pierre Trichet also ranked the viol immediately below the voice, remarking: "It must be acknowledged that after excellent human voices, there is nothing so charming as the tiny tremblings [*mignards tremblements*] that are made on the fingerboard, and nothing so ravishing as the dying strokes of the bow."[12]

Mersenne and Trichet illustrate how the voice—allied to God-given nature—offered an immutable standard against which to measure its instrumental imitators while at the same time affording scope for commentators to select those attributes on which vocal and instrumental technique best aligned. That is to say, what defined the voice depended on what instrument a commentator wished to praise. In the case of wind instruments like the recorder and cornett, as Jamie Savan has shown, articulation was central to claims of their unique proximity to the voice, varieties of tonguing being a means of imitating the pronunciation of speech.[13] In the case of the viol, the alignment was instead to be found in the continuous, breath-like control afforded by the bow. Likening the bow to the breath, Mersenne explained that the two shared both duration and expressive versatility: "For the bow which produces the effect of which we have spoken has a bowing almost long enough to be close to the ordinary breath of the voice, the joy of which it can imitate, the sadness, the agility, the sweetness, and the strength through its vivacity, its languor, its speed, its solace and succor."[14] By comparison with wind instruments, praising viols encouraged a less speech-oriented, more melody-and-dynamics-oriented conception of the voice.

The 1680s saw the first published music for solo viol, giving professional players the opportunity to repeat the viol's claim to being the most voice-like and hence best instrument. In the preface to his book of viol pieces, court composer-violist Le Sieur de Machy offered a succinct "panegyric to the viol" by observing, "The voice is the model for all Instruments and [the viola da gamba] is the one that imitates it best."[15] The pedagogue Jean Rousseau wrote more emphatically in his viol treatise: "One cannot dispute that an instrument has never come so close to resembling the human voice as the viol has, which indeed is only distinguished from it by the fact that it cannot actually articulate any words." He continued, "The viol knows only the voice above it."[16] Through such remarks, these sources by and for viol players perpetuated a hierarchical division of nature and artifice and a conception of musical instruments as asymptotically aspiring toward their model in nature.

And yet, with its fretted fingerboard and chordal capacities, the viol also had organological, playing-technical, stylistic, and repertorial connections to plucked string instruments, preeminently the lute. In the early to mid-seventeenth century, the viol's lute-like properties coexisted with claims such as Mersenne's and Trichet's. With the growing importance of solo viol playing in the 1680s, however, lute and voice came to figure as conflicting models. De Machy's *Pièces de Violle* (1685) reflects this emerging conflict, as well as his own commitment to the compatibility of a vocal ideal with lute-like techniques. Although de Machy repeated the maxim of voice as model for all instruments (as quoted above), this did not mean that the viol should play only in the manner of the voice. Instead, de Machy explained that the viol could be played in multiple styles, like the theorbo and harpsichord, and that the most proper style for the viol—as for all instruments that are to be played alone—was that of playing "Harmony Pieces."[17] To counter "those who want to argue that Solos of a single melody are preferable to those that are harmonious," he drew a comparison between playing a single melody on the viol and playing the harpsichord or organ with one hand: "This single playing might be very pleasant, but one would hardly call it playing the Harpsichord [or] the Organ."[18] According to de Machy, critics of "harmony pieces" wrongly believed that playing chords hampered playing a melody expressively with "beautiful graces" (i.e., ornamentation). In fact, one could play chords *and* "beautiful melodies with all the graces necessary for expressive playing." The solution, from a technique perspective, was to make use of two left-hand positions: one shared with lutists in which the thumb rests opposite the first finger and the fingers are each a half-step apart; and another in which the thumb is opposite the second finger, facilitating stretching out of the fingers to the alternating half and whole steps of a melody.[19]

Against the "bowed lute" style of playing, Rousseau argued for the vocal nature of the viol with a firm commitment to the voice as the one true model for the instrument. From the premise of voice as God-given model, he constructed a history for the viol. He rejected the idea that viol playing took its origin from plucked instruments on the grounds that these instruments' character was very different and "much inferior," owing to their lack of continuous sound and hence further remove from the voice.[20] Rousseau additionally maintained that the viol was the best instrument not only in his own day, but also always and originally. Thus, he reasoned that "since the viol is the most perfect of all, because it approaches closer to the natural than any other, we can judge that if Adam had wanted to make an Instrument, he would have made a Viol." He also spent several pages demonstrating that Orpheus's legendary instrument was not a plucked lyre but in fact a bowed viol, for which the ancients simply lacked the name.[21]

The idea that Orpheus's instrument was a viol also appears in earlier seventeenth-century sources. One example comes from a treatise on me-

FIGURE 2.1. Grotto of Orpheus depicting Orpheus playing a viol, Salomon de Caus, *Les raisons des forces mouvantes avec diverses machines tant utiles que plaisantes*, book 2 (Frankfurt: Jan Norton, 1615). Courtesy of the Smithsonian Libraries and Archives, https://doi.org/10.5479/sil.375615.39088008194136.

chanical and hydraulic automata by the garden engineer Salomon de Caus, published in 1615. De Caus's designs include a grotto of Orpheus, where the fabled musician is depicted in the classic scene of charming an audience of diverse animals with his music. De Caus did not name Orpheus's instrument in his text, but in the accompanying illustration it is clear from the shape of the instrument, as well as the hand positions on the bow and fret board, that Orpheus is playing a viola da gamba (not a cello, as often stated by modern commentators; see fig. 2.1).[22] De Caus proposed mechanics to move Orpheus's bowing arm, so that it would appear the figure played while the sound came from a hydraulic organ behind the scene.[23] Would these organ sounds have favored perceiving Orpheus's power over his listeners as derived from his singing voice or from the harmony of his instrument? If anything like the musical example de Caus included in his explanation of mechanical organs, the organ would have been programmed to play melody and accompaniment (fig. 2.2; audio 12). There is

FIGURE 2.2. Conclusion to the madrigal "Chi fara fede al Cielo" by Alessandro Striggio, arranged by Peter Philips, illustrating the style of music pinned onto barrels and sounded by pipe organ in de Caus's mechanical gardens. Salomon de Caus, *Les Raisons des forces mouvantes*, book 1 (Frankfurt: Jan Norton, 1615). Courtesy of the Smithsonian Libraries and Archives, https://doi.org/10.5479/sil.375615.39088008194136.

no mention of Orpheus's mouth in the text, which appears to be closed, while the angle of his left wrist indicates a lute-like hand position suited to chordal playing.[24] Despite such support for an Orpheus who drew magical harmonies from strings, Rousseau found in Orpheus's viol a means to deny originary status or special power to plucked instruments and to reserve such status and power to the voice and its closest imitators.

Rousseau's insistence on a singular, pure origin also manifests in his discussion of hand positions. De Machy, Rousseau complained, claimed "there are *two* bearings of the hand necessary for the perfection in Viol Playing." Rousseau sought to establish, by contrast, that "there is only *one* way of bearing the hand in playing the Viol, and that this way is natural." Prior to taking up viol pedagogy, Rousseau had written a treatise on singing (*Méthode claire, certaine et facile pour apprendre à chanter la musique*), and he worked to introduce vocal ornaments and terminology into viol playing while eliminating chordal ornaments like *tenües*. According to Rousseau, *tenües*—where a left-hand finger remains held down on a string to allow the note to continue vibrating while bowing other strings—"stand in the way" of achieving "the beauty of melody and its *agrémens*."[25]

The Prelude no. 4 from Marin Marais's first book of viol pieces, published in 1686, offers an opportunity to observe the *tenües* in action, alongside numerous melody-oriented ornaments (including the tremblement, a trill indicated by a comma; a mordant indicated by an x; a vibrato indicated with a vertical squiggle; and a slide indicated by a v-like connector between two pitches). The *tenües* occur in measures 15–18, where horizontal brackets indicate the note to be held down so it can resonate (where the bracket starts) and when to let it go (where the bracket ends). There is indeed a departure from the prevailing melodic style in measures 17–18; here, large leaps make use of string crossings to trace two descending lines a sixth apart. The result is less continuity of sound in the upper melodic line, as it alternates with bass notes. And yet, indicating *tenües* on the upper pitches in this passage, Marais uses the ornament to let the melody notes ring, suggesting he heard *tenües* not as "stand[ing] in the way" of "the beauty of melody" but rather as sustaining that beauty in another way (fig. 2.3; audio 13).

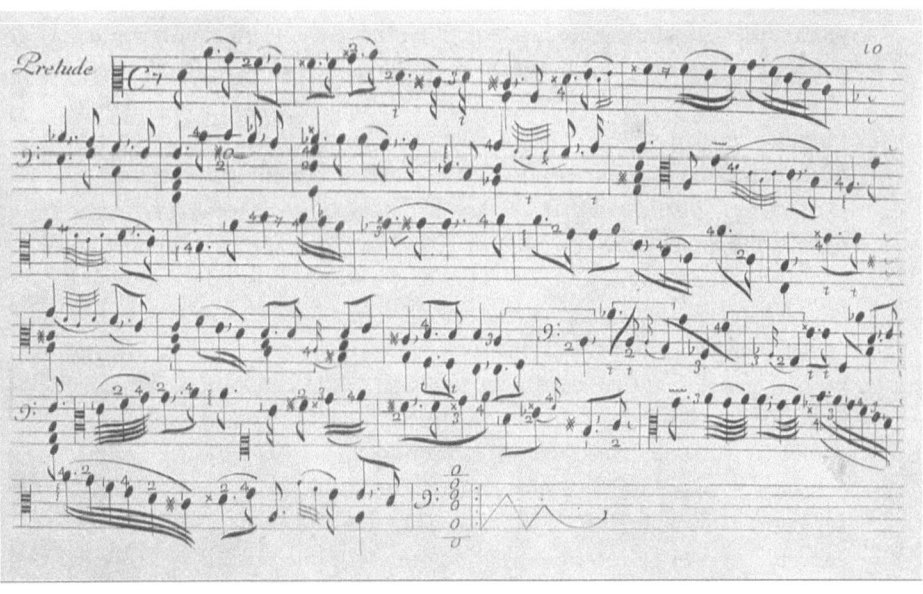

FIGURE 2.3. *Tenües*, indicated by horizontal brackets, call for letting notes ring in Prelude No. 4 from Marin Marais, *Pièces de violes . . . [livre 1]* (Paris: Auteur, 1686), 10. Bibliothèque nationale de France, ex-libris Descasaux, RES-756. Source: http://gallica.bnf.fr.

LISTENING TO A SOUND RING

While Rousseau placed a debt to plucked strings under attack in musical practice, it was newly emerging in studies of acoustics and music theory. The same year Marais published his first book of viol pieces, Joseph Sauveur became a lecturer and professor at the Collège Royal in Paris. Initially a mathematician, he turned to studying musical sound in the 1690s and began his first series of lectures on the topic at the Royal Academy of Sciences in 1696. These lectures concerned Sauveur's "new system of music," within which the concept of *sons harmoniques*—frequencies in whole-number ratios to a fundamental—was central. In the unpublished manuscript that was the basis for these lectures, Sauveur observed that "string instruments [and] bells contain these first intervals simultaneously when they are in good condition; that is, besides the principal sound, the octave, fifth, third, etc. are heard."[26]

Sauveur's subsequent publication, *Mémoires* (1701), began with a call for a new science—to be called *acoustics*—which "has as its object sound in general, whereas music has as its object sound in so far as it is agreeable to the hearing." Sauveur then proceeded to explain why he had not published his earlier attempt at acoustics (namely, the lectures he presented several years earlier). His reasons focused on things he had not sufficiently worked out at that time. These included the perplexities that arose from the observation that—under the right conditions—multiple harmonic sounds could be heard in a single tone: "While I was meditating on the phenomena of sounds, it came to my attention that, especially at night, you can hear in long strings, in addition to the principal sound, other little sounds (*petits sons*) at the twelfth and seventeenth of that sound, and that Trumpets, in addition to these sounds, make others, the number of vibrations of which is a multiple of the number of those of the fundamental."[27] He had found no satisfactory explanation for this phenomenon but, on further investigation, "concluded that a string, in addition to the undulations made by its entire length producing the fundamental, divides into two, three, four, and so forth, equal divisions, which produce the octave, twelfth, and fifteenth of that sound."[28] By observing vibrating strings (with the help of paper riders to make the vibrations more visible), Sauveur arrived at the concept of nodes to describe the still points that divide a string into harmonic subdivisions.

While strings provided the foundation for Sauveur's observations on harmonics, he readily generalized to other resonant bodies. Across his writings, he consistently paired strings with bells—just as he had in his first lecture series—for the way they rendered harmonics perceptible simultaneously in the same sound. In the *Mémoires*, he wrote, "Experience shows that long strings, when good and harmonious, render the first harmonics, principally those which are not in the same octave. Bells and other resonant and harmonious bodies make the same effect."[29] In his treatise on pipe organs (1702), he observed that organs imitate, through the mixture of stops, "the harmony which nature observes in resonant bodies, which are called harmonious; for we distinguish there the harmonics 1, 2, 3, 4, 5, 6, as in bells and, at night, the long strings of the harpsichord."[30] His treatise titled *Rapport des sons des cordes d'instruments de musique, aux flêches de cords* (1713) was concerned mainly with deriving the frequency of a string from its mass, length, and tension but also included a section on bells. Sauveur's discussion of bells was based on information he obtained from the Honorary Canon of Notre-Dame, Claude Chastelaine. From Chastelaine, Sauveur learned not only what pitches the bells were at churches like Notre Dame but also the bells' names. He listed these bells by their name and the tone provided by Chastelaine, followed by his conversion of its tone into his system of "fixed tones," which sought to standardize pitches by frequency. The reader is thus introduced to Notre Dame's Emanuel, Marie, Gabrielle, Guillaume, Pasquier, Thibaut, Jean, Claude, and Nicolas (these make up a minor scale plus a step, though Sauveur notes that the interval between Gabrielle and Guillaume is 1¼ tone); to St. Germain de Prez's Germain and Vincent (a whole step apart, with Vincent sounding the same tone as Guillaume); and more.[31]

Sauveur's writings establish the paradigmatic scene of listening for harmonics—or *petits sons*, as he called them, following Mersenne—as one in which the sound is produced by plucking or striking; it is in the dying reverberations that follow an initial impulse that one first discerns the *petits sons* ringing out. His repeated specification of listening "at night" to long strings suggests the especially still and quiet conditions, the close and attentive listening necessary to discern the *petits sons*. The one occasion on which he discussed actual bells, by contrast, suggests a very different listening situation—one in which great bells sound out across a city, their tones ringing from individually named objects as the voice of the church. Examining bell sounds in nineteenth-century France, Alexandra Kieffer has found a dramatic difference between poetic or religious and scientific discourses, the former attuned to the harmoniousness of bells, the latter to their inharmonic partials.[32] At the birth of acoustics as a science of sound, Sauveur understood bells as quintessentially harmonious bodies while simultaneously developing the string-based theories that would place bells outside harmony (their sounds "not... musical tones at all," in Hermann von Helmholtz's estimation).[33]

Composer-theorist Jean-Philippe Rameau took up the task of deriving the rules of music from a single principle in his first treatise, *Traité de l'Harmonie* (1722), but not until after its publication was he prompted to consider Sauveur's findings on harmonics as a resource in this endeavor.[34] In *Traité de l'Harmonie*, Rameau endorsed a Cartesian skepticism of sensory evidence, noting, "Conclusions drawn from experience are often false, or at least leave us with doubts that only reason can dispel."[35] Tuning his ear to harmonic overtones, however, he took an empirical turn, finding sensory evidence that dispelled all doubt about the validity of his music theory. He also found this evidence in a sound source that had not featured in Sauveur's scenes of listening for harmonics: the voice. As Rameau explained in the preface to his *Nouveau système du musique theorique* (1726):

> There is actually in us a germ of harmony which apparently has not been noticed until now. It is nonetheless easily perceived in a string or a pipe, etc. whose resonance produces three different sounds at once (this experience has been cited by various authors). Supposing then the same effect in all sonorous bodies [*corps sonores*], one should consequently suppose it in the sound of our voice, even if it is not evident. But to be more sure in this matter, I myself essayed the experiment, and suggested it to several musicians who—like myself—distinguished these three different tones within one tone of their voices. As a result, from then on, I have not doubted for a moment that this was the true principle of a Fundamental Bass, which I would have discovered again from this experience alone.[36]

From an initial note of caution about the perceptibility of harmonics in the voice—a recognition that previous authors had not observed them there—Rameau proceeded to confident confirmation of their presence with his own ears and the verification of other musicians. Now, to support the claim that the fundamental bass "is innate to [musicians]," Rameau appealed not to the division of a string into harmonic ratios for proof (as he did in the 1722 *Traité*) but rather to the fact that "its principle subsists in their very own voices." Likewise, to argue that melody is derived from harmony, he asserted that melody "is suggested by a principle of harmony that is within us, and this principle is none other than the chord that we hear in a single sound of the voice."[37]

Scholars of Rameau have noted significant shifts in his argumentation—as Thomas Christensen describes, "at one time or another, Rameau cast his theory of the fundamental bass in the varied rhetorics of neoplatonism, Cartesian mechanism, Newtonian gravitation, Lockean sensationalism, and Malbranchian occasionalism"—even as he remained committed to the core endeavor of deriving music from a single principle, that principle being harmonic.[38] The voice was one site on which Rameau's shifting mode of knowledge production played out. In the *Demonstration du principe de l'harmonie* (1750), Rameau framed listening to the voice as a means of Cartesian deduction, explicitly citing Descartes's *Méthode* as model for his procedure: "I began by looking into myself. I tried singing much as a child might. I examined what took place in my mind and voice." By the time he published *Observations sur notre instinct pour la musique* four years later, however, Rameau had acquired a greater interest in sensationist theories, which downplayed the innate and prioritized the acquisition of ideas from sensory experience. This epistemological shift is reflected in Rameau's lessened reliance on a special status for the voice based on it being *in* the listener and his turn instead to object lists like Sauveur's with the voice included. Thus, he explained that the principle of humans' musical instinct exists "in the Harmony that results from the Resonance of every Sounding Body, such as from the sound of our voice, of a string, of a pipe, of a bell, etc."[39]

Rameau's *Observations sur notre instinct pour la musique* (1754) was also the first publication in which he responded to the criticisms of the philosopher Jean-Jacques Rousseau, published the previous year in the "Letter on French Music." According to Rousseau (who shared a special investment in vocal melody as well as last name with the previously discussed viol pedagogue), sequences of chords were "always lifeless when not animated by melody." In defending his theory against the melody-first arguments of Rousseau, Rameau recalled the effort it took to first perceive harmonic overtones in the voice and put forward the bell as a sonorous object in which even an unpracticed ear could recognize the harmonic instinct at the basis of music. As he argued in his *Code de musique pratique* (1760), "Harmony, in its primitive & natural state, as given by sonorous bodies, of which our voice is one, must produce on us, which are passively harmonic bodies, the most natural effect, and therefore the most common to all. Hence it is that whoever, for lack of an exercised ear, is not very sensitive to the different successions of harmony, is at least instinctively sensitive to the sound of a perfectly sonorous body, like a beautiful bell."[40] In addition to Rousseau's criticisms of his theory deriving music from principles of harmony, Rameau had by this time also read dispatches from China by Jesuit missionary Jean-Joseph Marie Amiot, which included Amiot's contention that "dull organs" were to blame for Chinese listeners not appreciating European harmony. As Zhuqing (Lester) Hu has demonstrated, the issue of the ear's (in)sensitivity would provide a means for European philosophers to contort cultural difference into biological inferiority, rationalizing not just how certain individuals but also entire populations could fail to confirm the supposedly natural and universal laws of music.[41] In Rameau's parsing of natural instinct from an "exercised ear," the ringing bell becomes even more exemplary than the voice of musical sound, of sounding human (humans also being passively harmonic bodies); already, however, the bell was on its way to sounding the opposite.

DEFENSE OF THE VIOL

In the 1680s, when de Machy and Jean Rousseau were writing, viol players could confidently claim theirs the "first of the Instruments." In the early eighteenth century, however, musical styles and instruments from Italy found increasing favor in France, and as public concert life expanded, the violin family began to outshine viols with their sound and repertoire. As one observer wrote in 1738, the viol had been "very fashionable, and extremely cultivated," its playing brought to a state of perfection by the celebrated Marais; "but as the taste for Italian music has grown, the viol has been much neglected, because it does not give much sound, and because one can hardly hear it at all in Concerts."[42]

It was in this context that Le Blanc, an abbé and doctor of law, published a treatise aimed at saving the viol from obsolescence. He called it *Defense of the bass viol against the enterprises of the violin and pretensions of the cello* (*Defense de la basse de viole contre les enterprises du violon et les pretentions du violoncel*, 1740). *Defense of the bass viol* was one of two publications Le Blanc dedicated to Jean-Frederic Phelipeaux, Count of Maurepas, who served as minister of state to Louis XV from 1723 to 1749. No specifics of Le Blanc's relationship to Maurepas are known, but the minister's surviving papers suggest that his "chief concern in the 1730s and 1740s was to provide ... plans for the defense of France's far-flung and inadequately garrisoned empire."[43] Le Blanc's dedication to Maurepas noted that he was putting "the interests of the viola da gamba in your hands" and addressing "to you the defense of its rights," adding the musical instrument to the world of French subjects whose protection fell under Maurepas's responsibilities.

Le Blanc's *Defense* is an entertaining text. Parts are presented in Le Blanc's authorial voice narrating events, while others take the form of dialogue between personified musical objects, such as Musica, Sultan Violin, and Dame Viol. Ideas are often articulated through allusions to mythological or historical scenarios (presuming the reader's extensive knowledge of classical sources), and musical qualities are explained through such cross-sensory analogies as to the taste of wines or the shape of a lady's leg. Women have a significant presence as musicians, listeners, and arbiters of taste. These features of the text show Le Blanc to be both describing and addressing the fashionable world of court and salon.

To champion the viol, we might expect Le Blanc to draw on that standard panegyric—deployed, as we have seen, even by those who promoted a bowed lute style of playing—and make the case for it being the most voice-like instrument. The place of the voice in Rameau's *Nouveau système*, too, would suggest its special legitimizing power. Le Blanc, however, built his argument on different premises. Rather than deploying the voice as a singular ideal against which to measure the viol, Le Blanc distinguished between two categories of sound—voice and resonance—and then praised the viol for its union of the two: "The Viol holds the proper middle position, with its bow and its flexible strings; it gives to voice without taking away from resonance."[44]

Rebekah Ahrendt has connected Le Blanc's middle-ground argument to contemporaneous political theory and international relations: a portion of the *Defense* stages a diplomatic encounter between Sultan Violin from Italy and Lady Viol of France, and the viol's "happy medium between voice and resonance" exhibits the ideal of a balance of powers.[45] Le Blanc, whose musical tastes centered on repertoire from the first decade of the century, could also have drawn inspiration from other early eighteenth-century responses to the success of Italian music on French soil. François Couperin's *Les Goûts-réunis* (1724), for instance, exemplifies the bringing together of distinguishing features of the French and Italian styles, which Couperin characterized as having long divided "the Republic of Music."[46]

And yet, Le Blanc's account of the viol involves many more forms of relation than those between nations. Not only did he state that the viol "holds the proper middle position"; he also explained how the instrument attained this status in terms of it being "mongrelized" and becoming a "mulatto instrument." With these terms for mixed breeds and races, Le Blanc connected his understanding of the viol to the entangled concerns of eighteenth-century natural history and colonial empire (the latter being particularly salient for his dedicatee). As Emma Spary has observed, "it was in the study of the processes of change within living beings that *histoire naturelle* related most closely to the concerns of improving landowners, legislators, physicians, and many others involved with the centralized management of French institutions."[47] Le Blanc's thinking about the viol sits in company with both geopolitics and developing scientific theories about reproduction in living organisms.

"Do not be angry if I say you were a worm, or an egg, or even a kind of mud," Paris Academy of Sciences member Pierre-Louis Moreau de Maupertuis wrote near the start of his treatise on reproduction in living organisms, *Venus physique* (1745). Thus addressing his text to a fashionable feminine audience, a lady of the salon, Maupertuis anticipated some discomfiture at his contention that before an embryo begins to acquire human features, it is "only formless matter."[48] This theory of generation directly challenged the prevailing preformationist theories of his day, which envisioned God having created all species and individuals at the beginning of the world. Consistent with a religiously sanctioned view of "static order and divine foresight," preformationist theories held that the germ of each child was contained within a single parent and awaited a trigger to begin development.[49] Among the inadequacies of such theories, according to Maupertuis, was their inability to account for hybrids—creatures that displayed distinctive attributes from each parent.[50]

The contrast between Le Blanc's account of the viol and Rousseau's mirrors the difference between these theories of generation and their respectively dynamic or static orders of nature. As we have seen, Rousseau set the viol within a fixed hierarchy of instruments, a static order in place since the beginning of time (hence his ability to reason the viol into the hands of Adam and Orpheus, based on its place in the instrumental hierarchy). Rousseau did allow that different times and places had viols with different numbers of strings or tunings. However, these were deviations from the form achieved in France—the six-string viol tuned by fourths except for the major third between its middle two strings—which was "the most perfect, because it is more suitable to imitate the voice than any other."[51] The voice as immutable standard thus permitted recognizing the true, "most perfect" form of the instrument. Le Blanc's story of the viol, by contrast, was not one of perfection by more closely approaching a vocal model. Rather, it was a story of transformation, achieved by mingling characteristics from distinct taxonomic categories.

What are the relevant taxonomic categories for thinking about the viol? Much of Le Blanc's treatise is concerned with working out such categories. The first section of the text focuses on explaining the respective merits and deficiencies of viol *pièces* and violin sonatas. To do this, he introduces two distinctions—that between poetry and prose and that between melody and harmony. According to Le Blanc, poetry, melody, and French *pièces* partake in the same character, which contrasts with that of prose, harmony, and Italian sonatas. Historically, Le Blanc proposes that poetry and *pièces* came first, beginning with hymns to God and developing around values of the striking and extraordinary. Prose and sonatas developed later and are more suitable for the "business of everyday life" and ordinary conversation.

At this point, Le Blanc observes that "sonatas have been adopted in place of *pièces* because their style is more humanizing" (*comme humanisant plus leur stile*).⁵² Le Blanc sets this "more humanizing" style of sonatas in contrast to the qualities of artifice and exaggeration associated with *pièces*. He singles out Michele Mascitti's sonatas for special approval, observing that "one never grows weary of playing over [his] second and third books." And yet, while Le Blanc has many words of praise for sonatas and complaints about *pièces*, this is not the final word on the subject. Rather, Le Blanc goes on to trace a process of transformation through which the "initial aversion which caused the substitution of sonatas ... for *pièces* was vanquished."⁵³

Le Blanc's use of *more humanizing* as a positive but not the only or ultimate positive reflects the different place of *the human* in his thought from the one it would later occupy. For Le Blanc, *more humanizing* describes a movement away from the extraordinary and poetic toward the everyday and prosaic. *Dehumanizing*, in this context, would not mean debasing the human but rather something more like superhumanizing. In fact, when *dehumanize* (*deshumaniser*) entered French, it carried this alternate valence. The word did not appear in the 1684 edition of the official *Dictionnaire universel*, but it did in the 1701 edition, with the definition "to strip man of his natural feelings." The entry ventured that the word was imitated from the Italian of Giovanni Battisa Guarini's play *Pastor Fido* (first published in 1590).⁵⁴ In the play, Silvio—who dismisses love as "effeminate and soft"—is cautioned that love is "a human thing," and if he would forswear it, he must "beware that by dehumanizing yourself, you do not become a fierce beast rather than a God." As the first natively French example, the dictionary offers, "one must not dehumanize the man in favor of the hero."⁵⁵ Such usages cast *dehumanizing* as a transformation that, while sharing a loss of "natural feelings," could move in multiple directions toward many different kinds of nonhuman being—a range of possibilities that would subsequently be reduced to movement toward a condition of animal or machine.

To explain the transformation by which the "aversion ... for *pièces* was vanquished," Le Blanc shifted focus from the style of sonatas and *pièces* to the sound with which they are played. Previously, according to Le Blanc, the viol played *pièces* and employed "raised bow-strokes, all in the air, which resemble so much the plucking of the lute or the guitar." Le Blanc described the resulting sound onomatopoetically, with the consonant-ending "ticktock" (tic-tac). The violin's style of playing sonatas, by contrast, consisted in "drawing forth a continuous sound, which like the voice, is masterfully shaped in motion, like clay on the potter's wheel."⁵⁶ The celebrated viol player Marais, however, initiated a movement toward a more continuous sound on the viol. Through his compositions and playing style, Marais developed "a kind of harmonious sound, with the resonance of clock chimes" (*une nature de Son harmonieux, en résonnance de timbres de Pendule*).[57]

Though Le Blanc supplied no specific music examples to illustrate Marais's development of a "a kind of harmonious sound, with the resonance of clock chimes," we can get some idea of the sounds he describes by turning to Marais's second book of viol pieces (1701). Here, we find both a "Fantasie luthée" and a piece titled "Les cloches ou carillons" (see figs. 2.4 and 2.5). The former carried the instruction that the bow strokes should be "very short to better imitate the style of the lute."[58] The primarily monophonic line—arpeggiating chords rather than striking them as simultaneities—and dotted rhythms are also characteristic of lute playing; the lute teacher Perrine described both in his discussion of the "particular manner of playing all kinds of pieces for lute" in his book for lute players published in 1680.[59] Surprisingly, given the importance of the *tenüe* to the "bowed lute" style of viol playing, there are no specific indications to use this technique in the "Fantasie luthée." Together with the instruction to play with very short bow strokes, this suggests a perception of lute sound dying abruptly (like Le Blanc's "tic-tac") rather than ringing out (audio 14). The piece "Les cloches ou carillons," by contrast, is imitative of tower bells and calls for numerous *tenües*, including one held for two full measures. At another moment, the viol strikes a low A in octaves three times (the lower octave in a register rarely called for in the book of pieces and requiring a seven-string bass viol), and intervening rests supply the opportunity to hear the sound reverberate. The piece also makes prominent use of parallel thirds, and any note of longer duration than an eighth note is harmonized with some assortment of third, fifth, and octave above the lowest pitch. All of these features contribute to creating "a kind of harmonious sound" that rings out longer and with more harmonic richness than the plucked sound of the lute (audio 15).

Yet a further development in the viol's sound was to come by combining the viol with the violin's repertoire (sonatas) and bow technique. This took place when Antoine Forqueray, after hearing Marais, developed another way of playing "in which a sparkling sound [*un Son petillant*] resulted from a mature taste, reconciling the French harmony of resonance to the Italian melody of the voice [*conciliant l'Harmonie Françoise de résonnance à la Mélodie Italienne de la voix*]."[60] Le Blanc was particularly taken with the way Forqueray played the sonatas of Mascitti, a violinist who studied with Corelli before settling in Paris and seeking to reconcile (according to the

FIGURE 2.4. Viol imitating the short-plucked sounds of a lute in "Fantasie luthée" from Marin Marais, *Pièces de violes . . . [livre 2]* (Paris: Auteur, 1701), 43. Bibliothèque nationale de France, département Musique, VM7–6268 (A). Source: http://gallica.bnf.fr.

FIGURE 2.5. Viol imitating the long-resonating tones of a bell in "Cloches ou Carillon" from Marin Marais, *Pièces de violes . . . [livre 2]* (Paris: Auteur, 1701), 51. Source: http://gallica.bnf.fr.

FIGURE 2.6. A style of Italian violin music that combined with French viol playing to reconcile "voice" and "resonance," according to Hubert Le Blanc. Sonata Op. 2 No. 5, mvmt. 3 (Corrente Andante) from Michele Mascitti, *Sonate da Camera a violino solo col violone o cembalo* (Paris: Baussen, 1706), 20. Bibliothèque nationale de France, département Musique, VM7-706. Source: http://gallica.bnf.fr.

preface to his Op. 2) the Italian style with the "beautiful things" he had found in French music (fig. 2.6; audio 16).[61] But Le Blanc described not a compositional mixing of national *goûts*, but a mixture heard in the very sound of the instrument—a reconciliation of "voice" and "resonance."

It is in the context of Forqueray's efforts to keep viol playing alive that the term *mongrelize* appears in Le Blanc's text: "The true softening comes as a result of having mongrelized the instrument often, so that the sustained sound is rendered continuous like that of the voice" [*Le vrai adoucissement venant en consequence d'avoir souvent mâtiné l'Instrument, dont le son soutenu étant rendu continu, à la manière de celui de la voix*].[62] According to the *Dictionnaire de l'Académie française* (1740), *mâtiner* was said when a male dog mated with a female dog of "more noble species" (*plus noble espèce*), so one might say, "This ugly male dog has mongrelized this female hound" (*ce vilain chien a mâtiné cette levrette*).[63] The idea of *mongrelizing* the viol thus comported not only with Le Blanc's gendered personification of the viol but also with its original, "more noble" status as an instrument of the aristocracy, in contrast to the more public performance, general audience–oriented violin (as Donald Beecher aptly notes, debating what constituted the viol's *noblesse* was part of how coterie audiences enjoyed the instrument).[64] If the official dictionary betrayed anxieties about diluting noble blood, however, salon culture offered a different context. As Mary Terrall has noted, new dog breeds were a fashionable interest; breeding dogs was an activity in which salon women participated, and it played an important role in the development of theories of inheritance that challenged preformationism. In *mâtiné*, then, Le Blanc found a means to describe how the viol was transformed from an instrument destined for obsolescence to one able to reclaim an audience. Composers' interest in composing for it was restored, Le Blanc claims, and it regained the favor of listeners in intimate performance spaces.

After this favorable development in viol playing, in Le Blanc's telling, the "empire of the viola da gamba" was seemingly secured—and the violin decided to invade. To this end, the violin recruited the harpsichord and cello to join him in a mission to become "the three instruments which alone are necessary in music." Accepting the violin's proposal, the harpsichord and cello saluted him "as Attila, the scourge of the viola da gamba and as the exterminator of all the mulatto instruments [*Instrumens Mulâtres*]." With this, Le Blanc again applied a term for mixture in living beings to the viol, this time the term being for people of mixed European and African descent. In France's Caribbean Islands and lower French Louisiana, enslaved Africans outnumbered European colonists by the 1730s, and policies that were once permissive toward marriage and considered mulattos free citizens turned increasingly fearful and restrictive.[65] In French Louisiana, where the 1724 Code Noir prohibited marriage between Africans and Europeans, Jennifer Spear notes a "resounding silence" around African-European relationships, which stands in sharp contrast to the public discourse on Indian-European relationships.[66]

Le Blanc's choice of *mulatto* to describe the viol under threat by the violin stands out from the official positions prohibiting African-European sexual relations and the lack of public counterargument to such restrictions. It lends support to Chude Sokei's argument that "to even imagine such a coming together of oppositions such as human and machine requires a sociohistorical template to frame it" and that this template has been slavery.[67] Much of the rest of Le Blanc's treatise elaborates the opposition of *voice* and *resonance*, defining them through a gendered binary while also further describing their reconciliation in the sound of viol playing. While racial mixture supplied a template with which to conceptualize a mixture of sonic categories, a difference between Le Blanc and the nineteenth-century authors Chude Sokei discusses is that the former wrote before the transatlantic slave wars of the 1760s, which Vincent Brown has shown shaped the horror with which slaveholders increasingly imagined the threat of slave revolution.[68] That is to say, Le Blanc had not internalized "fear of reprisal, revolution, and racial retribution" as part of the slavery template.[69]

Introducing the concepts of "masculine Harmony" and "feminine Harmony," Le Blanc aligns these with voice and resonance, respectively, and explains them in terms of the physical properties of the objects that produce them. "Masculine Harmony" he describes as "resulting from [sounding] bodies which are hard to set in motion, without the vibration of strings or the tremblings of sections, from which resonance results after a stroke has been given. This Harmony draws on that of the voice." In "feminine harmonies," by contrast, "having less voice but being all resonance, the agitation of the pliant parts lasts for a long time."[70]

Voice (masculine harmony) and resonance (feminine harmony) thus provide a classification system that cuts across the more familiar organological taxonomic division into string, wind, and percussion instruments.[71] The organizing principle, rather than the mechanism of sound production, is the degree of rigidity or flexibility that determines an instrument's ability to vibrate passively, demanding more or less active effort by the player to maintain a sound. Instruments that have voice without resonance include the trompette marine, flageolet, transverse flute, and Regals (small tinkly bells that are "so destitute of resonance that they would have to be beat on unmercifully for any continuity of sound").[72] Instruments that have resonance without voice include the lute, theorbo, and recorder.

The categories of voice and resonance also have associated qualities of listening—a function of both the environment and the listener's attentiveness. Resonance is not appreciable in large performance spaces. It requires close-up listening and "an ear friendly to the trembling or vibrations of a sonorous body."[73] Resonance does not hit or strike the ear, but "caresses" it: "It is the principle of that [tender] harmony that it consists in the easy stirring of the parts of sonorous bodies which are set in motion by a light touch, from which there results at close hand a caressing of the ear by the quivering which causes the resonance."[74] As a result of its high string tension, by contrast, the violin relies on the open strings of the harpsichord to provide supplementary resonance. According to Lady Viol, the violin's sound is "only suitable to spare the curious the pain of listening attentively."[75]

Le Blanc's multiple classificatory systems come together in the idea of a sonorous note that combines voice and resonance, its continuity coming from free vibration rather than active drawing out. Thus, the violin, "through the extraordinary tension of the short, thick strings, sacrifices all to the voice." But the combination of bow (borrowing from the continuous style of the violin) and flexible strings (as opposed to the violin's high-tension ones) enables the viol to partake of both voice and resonance—exemplified in the playing of Marais and Forqueray when they "applied themselves to making the note sonorous (like the great bell of St. Germain) playing 'in the air' as they required, that is to say, having made the bow stroke they left a place for the vibration of the string."[76] That certain great bells, too, unite voice and resonance is a topic Le Blanc discusses at some length. The bells of St. German have "a vibration so charming that their Majesties, their Serenities and their Excellencies get up at night in order to hear them."[77] Le Blanc thus conjures a listening scene like the one Sauveur had specified for long harpsichord strings: night offers the stillness to better discern the harmonious vibrations in the sound. Le Blanc also notes that different bells sound different intervals with themselves (that is, in one and the same bell), including a fourth, a minor third, and a diminished fifth. Le Blanc's observations on bells help conjure the auditory experience of listening to a sonorous note; and they clarify that modern viol playing and great bells alike share the mixture of voice and resonance that he prizes.

A HYBRID VOICE

In 1700, Denis Dodart (physician to Louis XIV) presented his "Mémoire sur les causes de la voix de l'homme" to the Royal Academy of Sciences. He made detailed comparisons of the vocal apparatus to an oboe and organ stops. Having established the role of the reed in sound production and pipe length in pitch, he concluded that the mechanism of the voice is quite different, since the single opening of the glottis controls pitch by means of the "lips" (*lévres*) vibrating more and less and the different measurement of their opening. Thus, we should understand the voice as an "unknown species of Wind Instrument," "so ancient in nature, since it is as old as the human race, and yet so unknown in the Music of Wind Instruments, and so inimitable to the whole world of human industry."[78]

Dodart's conception of the voice as a physical instrument—but one that is unknown to and inimitable by human artifice—comports with the thinking exemplified by those whose who measured human-designed instruments against the voice, the perfection of which could only ever be asymptotically approached. In a commentary on Dodart's memoirs, Bernard le Bouyer de Fontenelle reiterated his point about the divinely created vocal instrument: since no wind instrument produced sound by means of varying a single opening, as the voice did with varying constrictions of the glottis, it seemed "that Nature had the design of placing [the instruments of the voice] altogether outside the realm of imitation."[79]

In 1741, Ferrein presented a new theory of the formation of the human voice to the Royal Academy of Sciences: "There are string instruments, such as the violin, the harpsichord; there are others of wind, like the flute, the organ. But we do not know any which are string and wind at the same time: this instrument... I found in the human body."[80] Ferrein thus theorized the voice as a novel instrumental hybrid—an entity one commentator described as "very paradoxical," though also well supported by Ferrein's experiments.[81] Ferrein explained, "I had found in the lips of the glottis strings that could vibrate and sound like the strings of a viol, and I regarded the air as the bow that sets them going, and the chest and lungs as the hand that draws the bow. I used this principle to explain the loudness of the voice, the variety of tones and many other phenomena the cause of which had previously eluded understanding."[82] The apparently large conceptual leap between the lips of the glottis and strings narrows in light of the tradition of comparing the viol to the voice. As we have seen, Mersenne likened the viol bow to "the breath of a human voice" in the length of its stroke and expressive means. The innovation of Ferrein's thought was not to liken voice and bowed string instrument, but rather to cross taxonomic categories to explain how the voice makes sound.

As with Le Blanc, Ferrein's hybridizing move involved some rethinking of what the key taxonomic categories were. Thus, while Ferrein invoked a familiar string versus wind instrument classification to announce the novelty of his discovery, he went on to revise this taxonomic system on a less mechanical, more acoustical basis. There are instruments that sound their material, and there are ones in which the material serves merely as a resonating cavity. In the former category, which includes bells as well as strings, "we hear the sound of gold, silver, copper, & c. of which they are made." In the latter, which includes flutes and whistles, "the materials used to make them, wood, lead, tin, ivory, are no more useful to the sound of these instruments than the forests and valleys to the sound of the echo." Ferrein concluded: "Silver rings in a bell, but not in a flute... these are admitted facts of all physicists."[83]

By abstracting from sound-producing mechanism to a relationship between materiality and sound quality, Ferrein brought bells and string instruments into the same taxonomic category. Further, with this rethinking of the taxonomic system, Ferrein argued that a reason the voice could not be only a wind instrument was that one hears its materiality. Ferrein summed up his demonstrations: "I had promised a wind and string instrument all at once, this commitment is fulfilled: we have just seen a pneumatic dichord more varied in its sounds and more harmonious than anything human industry could have imagined."[84] As we saw in seventeenth-century discourses, the voice as model for instruments meant highlighting qualities that were most successfully imitated, like articulation or note length and dynamic shape. Oriented instead toward explaining what set the divinely created instrument apart from other instruments, Ferrein cast the defining qualities of the voice in terms of its variety of sounds and superior "harmoniousness."

Ferrein's observation that "silver rings in a bell, but not in a flute" illuminates why it was *timbre* that became the word for tone quality. Historicizing the concept of timbre, Emily Dolan has noted the evolution of the term between the 1694 and 1762 editions of the *Dictionnaire de l'Acadmémie Françoise*. The earlier defined *timbre* as a kind of bell with a clapper inside that is struck by a hammer. The 1762 edition added an extended usage: "It is used sometimes figuratively for even the sound of the voice. And in this sense, one says of a beautiful voice: there's a beautiful timbre. This voice has a silver timbre."[85] That *timbre* could be used figuratively even for voices registers the connection the term still had to bells. A description of the singer Pierre Jélyotte's voice by Dufort de Cheverny, a participant in courtly life under Louis XV, reflects this connection and transference. Jélyotte sang leading male roles in most of Rameau's operas and inspired Graffigny's tears at the 1739 rehearsal of *Dardanus* discussed in the previous chapter. According to Cheverny, "his *timbre* was that of a perfect haute-contra, so full that certain sounds were as bright as if they came out of a silver bell."[86] Ferrein's observation suggests that, beyond being enabled by a perception of sonic resemblance between singing voice and silver bell (in the shared quality of brightness or brilliance in this case), the figurative use of *timbre* extended a way of perceiving bells as resonant of their materiality. That is, a mode of listening cultivated around bells—attuned to the "harmonious" sound ringing from a metal body—was generalized to listening situations where previously other modes of listening (such as Rousseau's attunement to melodic ornamentation) had predominated. These listening practices made *timbre* a ready term for the property of sound that is shaped by and identifies its material source.

Roseen Giles has suggested that changed perceptions of castrato voices mark an important paradigm shift in perceptions of timbre and materiality, this shift coinciding with the castrato becoming unnerving and the demise of its tradition. The cultural practices of listening to bells and theorizing sound, however, call for revising Giles's account of this paradigm shift as one in which "the castrato's voice changed from being natural to being metallic."[87] As Martha Feldman has shown, voices were described as metallic across the period of the castrato's celebrity; she reconstructs a vocal technique by which, she argues, singers produced the "ring" and

FIGURE 2.7. Jean-Jacques Rousseau's example of a consonant carillon, illustrating what he considered its lack of song, from his *Dictionnaire de musique*, Tome 1 (Paris: Veuve Duchesne, 1768), plate A. Bibliothèque nationale de France, département Arsenal, 4-S-3833. Source: http://gallica.bnf.fr.

"resonance" that metallic descriptors meant to capture. Thus, it was not so much the castrato's voice that changed status from natural to metallic. Rather, it was the metallic that changed status—from harmonious to inharmonious, malleable to unyielding—through such transformations, becoming the sound of a nonhuman ontological zone.

The transformation in the status of metal bells—their becoming distinctly nonhuman in opposition to the distinctly human voice—can be seen in the writings of Jean-Jacques Rousseau. Across the 1750s and 1760s, Rousseau developed the idea that music originated with the voice and a human desire to communicate emotions to one's fellow humans. It was from imitating human's instinctual vocal cries of passion that music derived its expressive power and ability to touch the soul. Catherine Kintzler has aptly characterized Rousseau's musical thought as a "counter-theory" because it was formulated in opposition to Rameau's harmony-centric theory. As Chua sums up Rousseau's disagreement with Rameau's instinct for harmony at the basis of music, "a harmonic humanity, because it is instrumental, is not human by definition."[88]

Bells, as we saw, were the sounding object to which Rameau turned in order to demonstrate the instinct for harmony in the wake of Rousseau's criticisms, for even an unpracticed ear could experience the power of harmony in an instinctive sensitivity to the sound of a beautiful bell. Rousseau heard bells differently. In his dictionary on music, Rousseau defined *carillon* as "a kind of air made to be performed by a number of bells tuned to different tones." The melodic possibilities in this kind of air were restricted by the length of time each bell rings; "all their sounds having some duration," Rousseau observed, the melody must be arranged "so that the sounds which continue together may make no dissonance to the ear." Rousseau illustrated the definition with an "example of a consonant carillon, composed to be executed on a clock with nine tones made by the famous clock-maker Mon. Romilly."[89] With its primarily triadic arpeggiations, prevalent octave leaps, and avoidance of stepwise motion, the musical example confirms Rousseau's prose description (fig. 2.7). But Rousseau goes beyond mere technical description, adding that composing a good carillon is "more tiresome than satisfying, for it is always a silly music, that of bells, even if all the sounds would be exactly in tune, which never happens."[90] Rousseau concludes: "It is obvious that the extreme constraints to which the harmonic combination of neighboring sounds and the small number of bells are subject does not permit putting any song in such an air."[91] Thus, Rousseau denies both any shared territory between singing voice and bells and any valid grounds on which to appreciate the sound of bells. Countering not only Rameau's theory of harmony but a culture of appreciative listening to sonorous ringing, Rousseau purified voice and bell into human and nonhuman.

DISAPPEARING HYBRIDS

Le Blanc and Ferrein both constructed alternate taxonomies, not to better fit their object of interest into a category but to better support their account of it *as a hybrid*. The fate of Ferrein's wind-string instrument is exemplary of an opposite tendency, in which hybrids—rather than drawing attention—disappear through the cracks of a taxonomic grid, either by being forgotten altogether or by persisting in "purified" form. As noted at the outset of this chapter, Ferrein's notion of *vocal cords* remains part of the basic vocabulary of vocal anatomy. Almost immediately, however, his hybrid conception of the voice fell out of the discourse. As Ferrein's student Claude Montagnat defended his teacher's theory against published critiques in the mid-1740s, he repeatedly addressed the issue of the voice's taxonomic identity. When the *Jugemens sur quelques Ouvrages nouveaux* reported that Ferrein found "the organ of the voice is not a wind instrument, as was thought, but a string instrument," Montagnat corrected this, noting, "The instrument of the voice is a string and a wind instrument at the same time."[92] To the anatomist Joseph Bertin—who premised his discussion of Ferrein on the question "is the glottis a string or a wind instrument?"— Montagnat wrote (with perhaps growing exasperation) "that M. Ferrein said expressly that it [the instrument of the voice] is a string and wind instrument at the same time."[93] Montagnat's efforts were to little avail; by the turn of the century, Ferrein was firmly identified with the view that the voice is a string instrument. In the popular physiology textbook *Nouveaux élémens de physiologie* (1801), Anthelme Richerand even seemed to be proposing a novel solution when he wrote, "Should one think, with Dodart, that the larynx is a wind instrument, or better to adopt the opinion of Ferrein, that regards it as a string instrument? . . . We ought to regard the larynx as filling at the same time the function of a wind and a string instrument."[94]

A similar fate befell viols. During the period of the violin's rising popularity in early eighteenth-century France, viol makers introduced instruments that combined viol and violin design features. One result was called the quinton, which Myrna Herzog has demonstrated was a five-string fretted instrument that appeared in France between 1725 and 1730. It had a violin shape and was tuned in fifths between the lower three strings and fourths between the upper ones. As Herzog also shows, twentieth-century scholars repeatedly attempted to categorize the quinton as either a viol or violin, with difficulties summed up by Sylvette Milliot's 1997 observation: "For a musicologist of our time it is rather difficult to define this hybrid instrument. Is it a viol or a violin?"[95] Herzog notes, "As a late French viol, its hybrid nature has been the main obstacle to its proper study," causing it to fall through the cracks of interest between violin players (to whom it was an unfamiliar instrument) and viol players (to whom it appeared as their instrument "contaminated" by violin features). Through her efforts to reconstruct the quinton, Herzog finds its sound to have "a characteristic ring, combining the resonance of the viol's high register with the solidity of the violin basses—a feature typical of late French viols."[96]

As makers of taxonomic categories and their mixtures, Le Blanc and Ferrein prefigure the "pleasure in the confusion of boundaries" and "responsibility in their construction" that Haraway called for in her "Manifesto for Cyborgs."[97] They also add support to Zakiyyah Iman Jackson's observation that "history would caution us against a quixotic celebration of hybridity."[98] In retrospect, their historical moment of enthusiasm for taxonomic borderlands helped set the conditions for enthusiasm for practices of purification. Le Blanc and Ferrein help us see why hybrids often appear initially as figures of new and transformative possibilities but in practice disappear—leaving powerfully naturalized distinctions in their wake. The pleasures of the hybrid—its status as a novelty, an advancement of knowledge, an aesthetic ideal—required and called into being the "pure" categories making it up.

THREE

Analogies
Diderot's Harpsichord & Oram's Machine

Speculation about how the brain functions and how to design a machine that simulates it, observed John McCarthy and Claude Shannon in 1956, "usually reflects in any period the characteristics of machines then in use." Thus, Descartes used "analogies drawn from water-clocks, fountains and mechanical devices common to the seventeenth century," early twentieth-century thinkers turned to telephone exchanges, and at the time of their own writing, it was "fashionable to compare the brain with large scale electronic computing machines."[1] McCarthy was a computer scientist who coined the term *artificial intelligence*, proposing that "every aspect of learning and any other feature of intelligence can in principle be so precisely described that a machine can be made to simulate it."[2] Shannon was a mathematician and electrical engineer who introduced foundational concepts for information theory. Both were leading figures in cybernetics, an emerging interdisciplinary field held together by core concepts like information, feedback, and control, construed to apply equally across living organisms and machines. With only slight variation, the three-phase history of technology was a common way of situating the conceptual and technical innovations of cybernetics. In Norbert Wiener's version, the age of clocks (seventeenth to early eighteenth century) was followed by the age of steam engines (the later eighteenth to nineteenth century), followed by his present-day age of communication and control.[3]

Notably sidelined by these cybernetic researchers and their followers are analogies to musical instruments, a conceptual resource used by Descartes as well as numerous other philosophers, anatomists, and engineers across the seventeenth to twentieth centuries. The absence of musical instruments from the cybernetic lineage can be chalked up to their effective alienation from the category of *machine*. In 1961, Barzun, a cultural historian, made telling use of that alienation in a preconcert lecture, delivered before the first public performance of works produced at the Columbia-Princeton Electronic Music Center. "You may be bent on proving that electronic music is not music," Barzun stated. "Most people of artistic taste share the widespread distrust and dislike of machinery and argue that anything pretending to be art cannot come out of a machine: art is the human product *par excellence*." Barzun offered a counterargument to such a view by reimagining musical instruments—accepted tools of art and human expression—as, in fact, machines: "Orpheus's lyre was a machine, a symphony orchestra is a regular factory for making artificial sounds, and a piano is the most appalling contrivance of levers and wires this side of the steam engine."[4]

McCarthy, Shannon, and Wiener linked a cybernetic discourse on organisms and machines to a particular, progressive lineage of machines. With its idea that each age is defined by one kind of leading technology, this lineage had the effect of cutting off the nascent field of artificial intelligence from a wider field of possible analogies through which to theorize and model how humans learn or think. A further effect of such restricted conceptual resources is to ease the slippage from analogy (keeping in view both resemblance and difference) to identity, from metaphorical to literal thinking about brains as computers. Such slippage happened quickly in the field of cybernetics. It was noticed four years into the establishment of the cybernetic community by neurophysiologist Ralph Gerard, who observed: "We started our discussion in the 'as if' spirit.... Then, rather sharply it seemed to me, we began to talk in an 'is' idiom. We were saying much the same things, but now saying them as if they were so."[5]

The analogical relations between organisms and machines developed by Diderot and Oram offer an alternative history and field of conceptual possibilities. A French Enlightenment philosopher and a British electronic musician, respectively, each drew on different musical materialities to explain sentient life. Compare a harpsichord and a philosopher, Diderot urged in *D'Alembert's Dream* (*Le Rêve de d'Alembert*, 1769). The harpsichord's keys are analogous to the philosopher's senses, its strings comparable to the nerve fibers of the body, which likewise oscillate and set one another resonating so as to enable feeling and thought. In sum, the philosopher is but a harpsichord endowed with sensibility and memory—a sentient instrument, or, as Diderot also put it, both musician and harpsichord at once. In her book *An Individual Note: Of Music, Sound, and Electronics* (1972), Oram explained how the graphical synthesizer she invented was designed to obtain *human* musical results from a machine. Elaborating what she meant by *human*, she developed multiple extended analogies between humans and electronic music equipment—explaining perception in terms of tuned circuits, the formation of individual consciousness in terms of formant control, thinking in terms of tape feedback, and more.

The significance of analogies for scientific thinking has received substantial attention from literary scholars and historians of science, particularly where they concern the operations of the mind. One benefit of attending to musical instrument analogies is their lesser familiarity, their strangeness compared to (for instance) computer analogies that have taken on a nonmetaphorical literalness—entering an "is idiom," as Gerard put it. *Dead metaphor* is the term for such figurative language that has lost its metaphorical quality, though Hayles prefers *dormant metaphor* to better capture the availability of the figurative dimension to becoming "active" once again.[6] To their "active" quality as metaphors for human functions, musical instruments add an additional dynamic, thanks to their oblique relation to the category of machine in the twentieth century. In other words, human-musical instrument analogies are not just human-machine analogies like any other, a kind of overlooked subcategory. Rather, human-musical instrument analogies throw sand into the gears of the modern notion of machine, disrupting the givenness of the category.

Diderot's image of a sentient harpsichord has been fairly widely discussed and has even appeared in writings on computer music as an anticipation of computers that store instructions (memory) and perform decision-making processes jointly with human players.[7] Oram's *An Individual Note*, by contrast, has as yet been little discussed, even since 2016, when a new edition brought the book out of obscurity. Bringing Diderot and Oram together provides a study in parallels and contrasts that serve to nuance the picture of how musical artifacts and practices relate to notions of *the human*. Though they drew on different sounding materials (such as strings vs. oscillators) and reacted to different problems (dualist philosophies vs. developments in computer music), they shared the image of human beings coupled to the world by vibration.

KEYBOARD ANALOGIES:
PIPE ORGAN, CARILLON, HARPSICHORD

By the time Diderot penned *D'Alembert's Dream*, analogies between humans and keyboard instruments were commonplace in philosophy and anatomical science. These analogies served to illuminate the workings of the human body—in particular, the mechanisms that translated external objects into internal sensations, or the connection between body and soul.[8] In his *Treatise on Man*, drafted in ca. 1632 and published in 1664, Descartes used a church organ to explain how the rational soul, located in the brain, received impressions. Descartes noted that an organist's fingers press certain keys and thereby cause air to pass from a wind-chest into certain pipes to produce harmony. Likewise, external objects stimulate certain nerves and thereby cause spirits to pass from cavities into certain pores in the brain to produce sensations in the soul. The organ player thus represented the external world, the pipe organ the human body, and a listener the soul that receives and makes sense of sensations produced by the body. A crucial lesson of this analogy, according to Descartes, was how to distinguish relevant from irrelevant factors in accounting for how the nervous "spirits" operate:

> The harmony of an organ does not depend on the externally visible arrangement of the pipes or on the shape of the wind-chests or other parts. It depends solely on three factors: the air which comes from the bellows, the pipes which make the sound, and the distribution of the air in the pipes. In just the same way, I would point out, the functions we are concerned with here do not depend at all on the external shape of the visible parts which anatomists distinguish in the substance of the brain, or on the shape of the brain's cavities, but solely on three factors: the spirits which come from the heart, the pores of the brain through which they pass, and the way in which the spirits are distributed in these pores.[9]

Even as Descartes's careful parsing of the analogy keeps the difference, the nonidentity between pipe organ and human clearly in view, the pipe organ also supplied a conducive model for imagining the nerves to be little tubes and the animal spirits to be a very subtle wind that travelled through them.

While he subscribed to a pneumatic model of nervous transmission, however, Descartes also believed the tubes of the nerves to contain delicate filaments or fibers akin to strings. These fibers ran from the sensory organs to the brain, and when an external object activated a fiber, it produced a slight tug that opened a pore in the brain. To explain this system, Descartes used another keyboard instrument analogy: the body was like a carillon. Again, external objects were like the player depressing the keys. In the carillon, however, depressing a key pulled a string, which activated a bell. In the case of the human body, the effect of pulling a string—of activating a nerve fiber—was to open a brain pore, producing sensation. The pipe organ and carillon thus supplied different models of nervous transmission—of the material connection between world and mind—

while offering fundamentally similar models for "the mind as a sovereign ruling over a polity of subservient body parts."[10]

In the mid-eighteenth century, vibrating strings superseded pneumatic tubes as leading physical analog for the nerves. The English physician David Hartley influentially developed a string-based neurophysiological model in his 1749 treatise *Observations on Man*.[11] Hartley argued that sensory impressions induce vibrations first in the nerves and then in the brain, where they survive in attenuated form to become simple ideas that can be associated with other ideas.[12] Though Hartley dismissed as absurd the notion that the nerves vibrate in the manner of "musical strings," he explained the behavior of ideas in the mind through extended musical analogies.[13] He suggested, for instance, that while the brain must be subject to a very complex set of vibrations, nonetheless—just as at a concert, "one instrument generally strikes the ear more than the rest"—one part of the brain's vibrations will prevail at any given time.[14]

Organ metaphors faded from use along with the tube theory of nerves, part of a broader shift in medical and philosophical thought from *mechanistic* to *vitalist*. Keyboard instruments, however, remained a primary model for explaining how external objects became internal sensations—for translating between outer and inner. Thus, to explain the passions, the abbé Armand-Pierre Jacquin suggested in 1762 that "it is good to consider our individual as some kind of musical instrument." The nerves, originating in the brain and spreading out "into an infinity of fibers... across the whole body," were "the keys and strings of the instrument. The exterior objects that strike the extremities of these fibers... are the more or less skillful hands that play the instrument. [And] the soul... is the ear of the one who listens to the instrument, and who is pleasantly gratified or cruelly torn."[15] As with Descartes, the material instrument provided Jacquin a metaphor for the body, while an immaterial listener completed the analogy to a feeling, thinking human.

Diderot's use of human-instrument metaphors spans his writings, and his early uses share with other instances of the metaphor a separation of body and soul. In his *Letter on the Deaf and Mute* (1751), Diderot took up the question of how the successive nature of speech relates to the simultaneous nature of experienced sensations and ideas. In providing an answer, he invoked a clock bell:

> Consider man as a walking clock... look on the head as a bell furnished with little hammers attached to an infinite number of threads which are carried to all corners of the clock-case. Fix upon the bell one of those

little figures with which we ornament the top of our clocks, and let it listen, like a musician who listens to see if his instrument is in tune: this little figure is the soul. If many of these little threads are pulled at once, the bell will be struck several times, and the little figure will hear several notes simultaneously... add that the sounds produced by the bell do not die away at once, but have some duration; that they produce chords with the sounds that follow, and the little figure that listens compares them, and pronounces them harmonious or dissonant; that memory, which we need to form opinions and to speak, is the resonance of the bell; the judgment, the formation of chords; and speech, a succession of chords.[16]

The strings of Diderot's clock are reminiscent of the strings of Descartes's carillon—they do not vibrate but rather convey a tug from one place to the another, thereby setting a bell in motion. Diderot adds to Descartes's carillon the duration of a bell's resonance as significant to the metaphor, accounting for memory and enabling judgement. Diderot also turns the listener into a musician at his instrument and suggests the threads dispersed to "all corners of the clock-case" are activated not by a single keyboard player but by a diffuse and impersonal world. Riskin has observed that Diderot does not specify whether the listening "soul" in this human-clock analogy should be understood to be material or immaterial.[17] Like Descartes's soul, however, it is situated in the head, perched atop and seemingly distinct from the nerves and resonating body. That the metaphor lent itself to a dualist understanding, separating material instrument from immaterial listener, seems apparent from the way Diderot revised his use of musical analogy in *D'Alembert's Dream*.

DIDEROT'S HARPSICHORD

D'Alembert's Dream is a text fancifully conceived as a dialogue between Diderot and the mathematician Jean D'Alembert, followed by D'Alembert's musings in his sleep. The dialogue begins in medias res, with D'Alembert saying that if one rejects the idea of an immaterial soul and replaces it, as Diderot suggests, with sensitivity as a "universal and essential quality of matter," one is led to the ludicrous notion that "stone must feel."[18] Diderot replies that the idea that a stone feels is hard to believe only for "the man who cuts the stone, carves, and grinds it without hearing it cry out." With this, Diderot challenged the connection between feeling and vocal expressivity and the uses to which such connection were put by his contemporaries. The vocal "cry" was a centerpiece of aesthetic theories that explained the capacity of music to arouse emotion through its relation to instinctive cries of feeling. Rousseau, for example, argued that vocal cries were the "first language of mankind, the most universal and vivid." The notion of the cry, in turn, grounded Rousseau's conception of music as an exclusively human language of emotion. Music does the most of any of the arts to "relate man to man," he argued; but that power requires a unique connection between musical sound and being human, a premise summed up in his claim "birds whistle; man alone sings."[19]

From the possibility of silent but nonetheless feeling stone, Diderot goes on to argue that there is both active and latent sensitivity, and it is this difference that distinguishes animate from inanimate matter. We should thus envision an object world that is latently or potentially sensitive rather than a world full of inaudibly screaming and sighing objects. Diderot's line of questioning puts him in company with Saint-Foix's Lucinde and Graffigny's Zilia (discussed in chap. 1), characters who modeled the pleasures and virtues of imagining sensibility in birds, statues, and people where others denied it. Rousseau exemplifies the impulse to narrow the sphere of sentimental sympathy; intent on defining the borders around "man," Rousseau devoted significant effort to clarifying what distinguished him from animals as well as what was natural to him as opposed to being a corruption by society. Diderot, by contrast, argued that "every animal is more or less a human being, every mineral more or less a plant, every plant is more or less an animal.... There is no quality which any being does not share in."[20]

And yet while sharing qualities, differences remain, and D'Alembert asks Diderot to explain the difference "between a man and a statue, between marble and flesh."[21] Diderot invokes active and latent energy, which D'Alembert understands. But whereas D'Alembert concurs that latent energy can transform into active energy, he does not see how something latently sensitive, like a marble statue, can become actively sensitive. Diderot replies that the conversion is common, taking place when minerals are absorbed by plants and plants are eaten by animals. The ability to convert inanimate matter into plant life and then to animal life satisfies D'Alembert on the point that all matter may be considered either latently or actively sensitive. Still, D'Alembert points out, "the sentient being is not quite the same thing as the thinking one."[22] An inert being may become sentient—that is, capable of feeling; but how, D'Alembert wants to know, does a sentient being become a thinking one?

Diderot's answer involves multiple parts. Firstly, he proposes that memory is all that is necessary for feeling to become thinking, for it is memory that allows linking together elements from one's life, forming a consciousness of oneself and performing such mental operations as denying, affirming, and drawing conclusions. Thus far, D'Alembert agrees. But he sees a problem: How is it possible to hold more than one thing in mind at the same time such that, for instance, the understanding can examine an object while also "concerning itself with affirming or denying certain qualities of the object"?[23] The question recalls the earlier context in which Diderot introduced the clock analogy, when—observing that the sound of a bell rings for some time—he proposed that memory was "the resonance of the bell," allowing more than one thing to be held in mind at once so as to be compared and judged.

In *D'Alembert's Dream*, Diderot pursues a different analogy; it is at this point that vibrating strings enter the picture. This very issue, Diderot tells D'Alembert, "has sometimes led me to compare the fibres of our organs with sensitive vibrating strings." The ongoing oscillations of a plucked string hold an object in mind while the understanding is busy considering a quality appropriate to it. Moreover, there is sympathetic resonance; by this action, one idea summons another, and these a third, and so on, with the result that there is no limit "to the ideas called up and linked together in a philosopher who meditates or listens to himself in silence and darkness."[24] With "listen[ing] to himself in silence and darkness," Diderot likens philosophical reflection to the nighttime listening we saw described by Sauveur and Le Blanc in the previous chapter, conducive to perceiving the harmonious resonance of a plucked string or bell.

In response, D'Alembert points out a flaw in the plucked string analogy, if Diderot indeed wants to maintain there is only matter and not spirit: "You make the philosopher's understanding an entity distinct from the instrument, a kind of musician listening to vibrating strings and making pronouncements about their harmony or dissonance."[25] In other words, Diderot has accounted for what the mental faculty of "understanding" operates upon in material terms but left the mental faculty itself unexplained—or rather, explained only as another person within a person's mind, like the "little figure who listens" in his clock metaphor. The comparison to vibrating strings thus amounts to the same kind of instrument analogy used by Descartes and the many others for whom it supported a dualist conception of the human.

"I may have laid myself open to that objection," Diderot replies. "But perhaps you would not have raised it if you'd considered the difference between the instrument called philosopher and the instrument called harpsichord. The philosopher-instrument is sensitive, being at one and the same time player and instrument."[26] To clarify how sentience is in the instrument rather than outside it, he elaborates the analogy: "Suppose that the harpsichord has both sensitivity and memory, and then tell me whether it won't know and repeat on its own the melodies you will play on its keys. We are instruments possessed of sensitivity and memory. Our senses are so many keys which are struck by things in nature around us, and often strike themselves. And there we have, in my judgment, everything which goes on in a harpsichord organized like you and me."[27]

The harpsichord has struck modern commenters as an odd choice of analogy for Diderot's philosophical instrument, due to its seemingly insensitive nature. Veit Erlmann, for example, observes the apparent incongruity between the "mechanistic inevitability" of the harpsichord and the image of the human as a vibrating bundle of nerves variably responsive to the outer world. The inexpressive fixity with which the keys of the harpsichord cause its strings to be plucked is captured by Couperin's remark that it "appeared almost impossible to... give any 'soul' to this instrument." Erlmann reconciles the incongruity by noting that a sense of "mechanical inevitability" remained "at the heart of Enlightenment notions of interplay of mind, body and environment"; he also suggests that the popularity of harpsichord analogies in general had to do with the instrument's "unique status as both cultural capital and a show piece of the bourgeois household, with its light, silvery sound being seen as emblematic of inner refinement and delicacy."[28]

Probing further the material specificity of harpsichords and their relationships to players and listeners, however, reveals a different picture of the instrument and what it contributed to Diderot's account of the human. When Couperin observed, in his 1716 treatise on the art of harpsichord playing, that "it has seemed almost impossible, up to the present, for anyone to give soul to this instrument," this was not a final word on the instrument but rather a context for the significance of his playing techniques. As Couperin explained, the sounds of the harpsichord were "determined" such that they "cannot be increased or decreased" (i.e., made louder or quieter); this meant that "giving soul" by way of dynamics—as would a violinist or flute player—was not possible. Couperin, however, achieved the "impression of feeling" (*impression-sensible*), successfully "touching persons of good taste," by means of timing the beginnings and endings of notes— either releasing them earlier than notated or initiating them later than notated. The early release he called "aspiration" (suggesting taking a breath), and the delayed attack he called "suspension" (suggesting waiting with expectation). While the harpsichord's sound was "determined," these modifications of time worked to "leave the ear undetermined." The result of the suspension (delayed attack), Couperin contended, was to produce "the same effect on listeners as when bowed instruments increase their tone."[29]

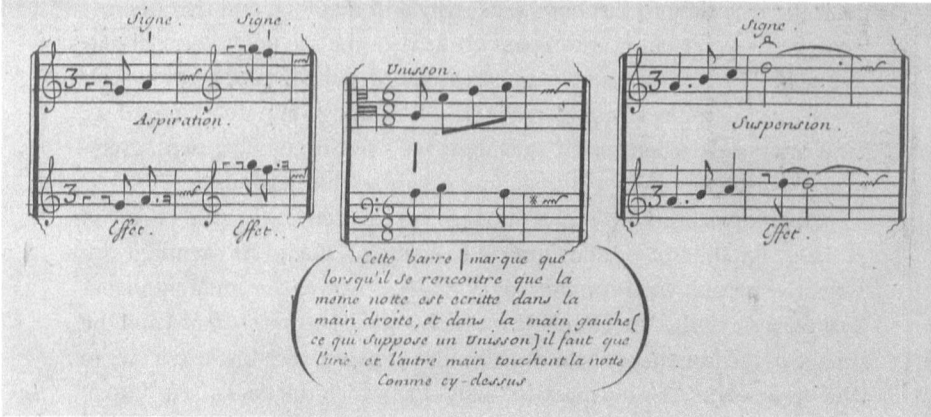

FIGURE 3.1. Explanation of signs for "aspiration" and "suspension," ornaments with which to "give soul" to the harpsichord, from François Couperin, *Pièces de Clavecin*, book 1 (Paris: le sieur Foucault, 1713), 75. Bibliothèque nationale de France, département Musique, RES F-76. Source: http://gallica.bnf.fr.

In his first book of harpsichord pieces, Couperin described the sound of the harpsichord as "perfect" and "brilliant" while reiterating that because its sounds could not be swelled or diminished, it required other techniques to "succeed in making this instrument susceptible to expression."[30] The book introduced signs for when to perform the aspiration and suspension, explaining how to perform the signs by providing the musical equivalent with notated rests (see fig. 3.1). The aspiration sign appears frequently throughout the pieces in the book, but the suspension is rare, appearing in only a handful of pieces. One of those pieces, "Les Sentimens Sarabande," includes the performance direction "very tenderly" (*tres tendrement*). The suspension is indicated three times (mm. 11, 13, 18); each time, it serves to delay a single right-hand melody note relative to the bass (fig. 3.2; audio 17). Taking seriously Couperin's notion of an "undetermined" ear, we can imagine the delayed melody note having started "on time" and then grown louder. With such a performance and listening practice, the harpsichord becomes exemplary of dynamic interaction—its music not a fixed, unidirectional product of struck keys but a coproduct between instrument and perceiver. The challenge of "giving soul" to the harpsichord is thus illuminating—not of how ill-suited the instrument was to modeling sentient, vital matter but of how conducive it was to entangling matter and mind.

We may identify further factors that made the harpsichord conducive to Diderot's theory of the "philosophical instrument" by considering who played the harpsichord. In Diderot's world, while men played and composed for the harpsichord as professional musicians, the instrument's players were predominantly young women. According to the *Encyclopédie* that Diderot edited, it was typical for a young woman to spend "three quarters of the day before a mirror or a harpsichord."[31] Diderot's own daughter, Angélique, started learning to play the harpsichord by the age of nine, and Diderot seems often to have been present during her lessons. In 1762, Diderot wrote to Sophie Volland about how delighted he was with his daughter, because she "reasons about everything she does" (*raisonne tout ce qu'elle fait*). To exemplify, he related the following exchange between Angélique and her harpsichord teacher: "Angelique, this passage confuses

FIGURE 3.2. Uses of the "suspension" to create the effect of a melody note growing louder in "Les Sentimens Sarabande" from Couperin, *Pièces de Clavecin*, book 1 (Paris: le sieur Foucault, 1713), 11. Source: http://gallica.bnf.fr.

you? Look at your paper." "The fingering is not written on my paper, and this is what stops me." "Angelique, I think you skip a measure." "How could I skip it, when I still hold the chord under my fingers?" These exchanges showed Angelique reasoning and asserting herself rather than passively following her teacher's instruction. This is what Diderot praised in his letter; but he immediately turned elegiac, resigned to the fact that such female agency was not to be nurtured: "What a pity that education responds so poorly to natural talents! The pretty woman that this would be one day!"[32]

The expected fate of women as opposed to men is neatly captured in the *Encylopédie* entries on these two topics. Man is defined as "a feeling, reflecting, thinking being, who freely walks the earth, who seems to be at the head of all other animals whom he dominates."[33] By contrast, we read of "woman": "This name alone touches the soul, but it does not always elevate it; it only gives rise to pleasing notions, which become by turn unsettling sensations or tender sentiments a moment later; and the philosopher who thinks of contemplating it is too soon only a man who desires, or a lover who dreams."[34] *Woman* thus figures as an external object, a stimulus for a "feeling, reflecting, thinking being"—a *man* whose thinking and reflecting are put at risk by the feelings aroused by *woman*.

Such gender dynamics contribute to the logic of Diderot's sentient harpsichord. The image of a woman playing the harpsichord morphs easily into a woman "playing" the thoughts of a philosopher. When Diderot writes "our senses are so many keys which are struck by nature," he maps the harpsichord onto a male body—an implicit gendering that becomes explicit when D'Alembert replies that if this sentient and animated harpsichord were additionally endowed with the faculties of feeding and reproducing itself, it would "either on its own or with its *female* partner, give birth to little keyboards, living and resonating." The idea that a harpsichord might reproduce "on its own," meanwhile, points in another direction: to the polyp (that plant-animal creature discussed in the previous chapter) and the possibility that harpsichords endowed with such vital capacities might reproduce asexually rather than sexually.

The radical implication of the polyp—that the soul or vital principle is not unitary and sovereignly seated but rather divisible and motile—was one Diderot found appealing and made use of in other musical moments. After Diderot took two friends of J. C. Bach to hear Marie-Emmanuelle Bayon play the harpsichord in 1768, he recounted in a letter: "She played like an angel. Her soul was entirely at the ends of her fingers."[35] Diderot developed the idea of a mobile embodied soul, suggested by this phrase, in his earlier novel, *The Indiscreet Jewels* (1748). The erotic novel includes a conversation between the Sultan Mangogul and Sultana Mirzoza on the topic of metaphysics. Mangogul espouses the traditional view that "the head thinks, imagines, reflects, judges, disposes, and commands." Mirzoza proposes an alternative theory, based on empirical observation, according to which the soul starts in the feet (evidenced by kicking in the womb) and then moves upward to other parts of the body as a person ages and develops. Rather than arriving at a final destination, however, the soul "moves along, travels, leaves one spot, comes back to it only to leave it again; but I maintain that the other limbs are always subordinated to the one in which the soul resides. This varies according to age, temperament, and circumstance. From there, differences of taste, diversities of inclinations and characters are born."[36] More than a poetic turn of phrase, then, Diderot's observation that Bayon's soul was "entirely at the ends of her fingers" as she played the harpsichord represented further empirical evidence for the kind of materialist theory explained by Mirzoza. At the end of their debate, rather than have Mirzoza win, Diderot satirizes the chauvinism of Mangogul, expressed by casting certain people as mere machines: "far from agreeing with you that every creature that has legs, arms, hands, eyes, and ears like me must also have a soul like me, I tell you that I am thoroughly persuaded that three-quarters of men and all women are no more than automatons."[37]

Lastly, we can add decorative mottoes and paintings to what made the harpsichord a suitable model for the philosopher. Roger Moseley has interpreted mottoes—typically found above the keywell—as part of a tradition of granting agency to instruments.[38] For as Thomas McGeary has shown, many mottoes invite us to imagine that they are "spoken" by the instrument itself. Consider the Latin motto *Dum vixi tacui mortua dulce cano*, which appeared on a number of harpsichords and translates as, "While living I was silent; now dead I sing sweetly."[39] According to Sheridan Germann, such mottoes explain some of the imagery commonly found on French soundboards, like that of a bird perched on a dead and broken tree stump that is nonetheless sprouting green leaves. The message of motto and image alike is *from dead matter comes new life*.[40]

Yet, while first-person mottoes certainly encourage us to imagine a harpsichord subjectivity, it is less clear how far they go toward marking an instrument's agency or autonomy. Consider an instrument originally built by Joannes Couchet around 1680, rebuilt with an added second manual by Pascal Taskin in 1768 and with a Latin motto on its keywell: *Non nisi mota cano*—"Not unless moved do I sing."[41] The same would seem to hold for

Diderot's sentient harpsichord; it is a feeling, reasoning yet passive instrument, a bundle of reactions to its environment. Or rather, its activity (thinking, feeling, and otherwise) is constituted by its material organization and responsivity to its environment, rather than being directed by a separate, immaterial agency.

It is harder to envision *man* as a "being, who freely walks the earth, who seems to be at the head of all other animals whom he dominates," if he is organized like a harpsichord in this sense. And indeed, that man holds no such special dominion is a conclusion Diderot is happy to draw—or rather, to *describe*. For when D'Alembert notes that the harpsichord model fails to explain how we draw conclusions, Diderot replies: "But we don't draw them at all. They are all drawn by nature. All we do is describe connected phenomena whose connection is either necessary or contingent, phenomena which we have learned by experience."[42]

In his earlier, 1751 *Letter*, immediately after Diderot drew the analogy to a clock with bells to explain the simultaneity of sensations, he remarked, "But enough of this language of metaphor, which at best is but fitted to amuse and arrest the volatile mind of a child; let us come back to philosophy, which requires arguments and not analogies."[43] Metaphorical language, indeed, had become a target of suspicion among Enlightenment thinkers, whose demotion of analogy to rhetorical flourish or tool for the weak-minded was part of what distinguished their knowledge claims from those of esoteric traditions. In *D'Alembert's Dream*, however, Diderot turned analogy on analogy itself, explaining that "an analogy in the most complex cases is only a three-part rule which takes place in the sensing instrument"; given a relationship between two phenomena, extrapolating the same relationship from another phenomenon is "a fourth harmonic string, proportional to three others." The turn realizes what Jess Keiser has argued more generally—that when "eighteenth-century thinkers pushed back against the excesses" of certain metaphorical accounts of the thinking, feeling human body, "they did so by writing different ones."[44] If there is a historical path toward increased understanding of the brain, this suggests, it does not take the form of a sequence of metaphors approaching literal truth but rather of multiplying analogies.

"PERHAPS MORE HUMAN"

Like Couperin at the harpsichord, Oram was concerned with how to impart human qualities or properties to a musical medium that seemed resistant to them. In the early 1960s, having departed the BBC Radiophonic Workshop to set up an independent studio and lead the way in British electronic music, Oram formulated her driving concern as being how to make electronic music with human feeling. Achieving this would involve a focus on two things generally being overlooked in the international scene of electronic composition, as she saw it: mind and hands. Central to her advocacy of electronic sound for music composition was a call to use it to compose within—as well as to systematically test—the limits of human comprehension. She was also committed to the idea of hand drawing as carrier of composerly thought and feeling. As Oram engaged with developments in computer arts and cybernetic discourses in the late 1960s, however, she developed a more concerted focus on defining "what makes us human" and how to "humanize the machines." Tracing how Oram's interventions around human and machine changed from the early 1960s to early 1970s illuminates her own views and the shifting landscape in which she formulated them, as well as the formative role played by analogies.

How Oram situated herself vis-à-vis the state of electronic music around 1960 is well illustrated by a program of electronic music she presented at the Edinburgh International Festival in September 1961. Adopting the format of an illustrated lecture, she explained techniques that had developed around magnetic tape machines and tone generators—techniques like tape editing, mixing sine waves, and filtering noise. She also played compositions that demonstrated how composers were using these techniques, including: Herbert Eimert manipulating sounds produced by electronic generators in *Fünf Stücke*, no. 4 (Cologne Radio's Studio for Electronic Music, 1955–1956); Niccolò Castiglioni mixing electronic sounds with "natural sounds" (i.e., sounds recorded via microphone) in *Divertimento* (1961); and Włodzimierz Kotoński manipulating natural sound only in *Study on One Cymbal Stroke* (Polish Radio's Experimental Studio, 1959).[45] Making such electronic music, Oram noted, composers no longer had the sonic constraints of vocal cords, ten fingers, and orchestral instruments. But she cautioned, "We've got to beware on this that we stay within the comprehension of the mind, and I would like to emphasize this throughout this evening. That we must—although we seem to have thrown overboard many of the limiting factors—we have still to consider that this must be within the comprehension of the mind." Many composers had transferred serial techniques from traditional to electronic composition, thereby finding a means to create compositional "rules" for the new medium that in itself offered "virtually none." These rules, however, were not grounded in a concern to "stay within the comprehension of the mind."[46]

The music Oram included of her own composition demonstrated how one could instead start from "the comprehension of the mind." After discussions of tone color, pitch, and rhythm, Oram introduced "one last thing:

volume." Contrary to what one might think, making a crescendo was not just a matter of turning up the volume knob: "To get a real feeling of crescendo one has got to increase the color. Perhaps put more and more high frequencies into the tone color as well."[47] Oram exemplified this principle with an excerpt from her piece "Four Aspects" (1960, audio 18). Starting around 2:16 in this composition, a low hum begins to take over, gradually growing louder and being joined by higher frequencies (especially around 2:40) until coming to an abrupt halt around 3:30. On another occasion, Oram explained that this "experimental piece" used "the simplest of musical form—A.B.A. coda" and "mostly the simplest of harmonies (tonic/dominant)," but "every sound in it is electronic." Thus, rather than transfer serial technique, Oram transferred basic structures and harmonies from tonal music. By combining novel sounds with simple structure and harmony, she constrained her compositional task to "facing only one new problem at a time," which, as an approach to the vast new realm of electronic music, served to "educate ourselves to one new facet at a time."[48]

The question-and-answer period that followed Oram's presentation reveals what a challenge it was to make sense of, or find pleasure in, the compositions she played that evening.[49] The audience wondered: Has the composer a definite plan? Is musical training even necessary to produce this music? What criteria does one use to distinguish a bad from a good piece? Is there really a lot of accident involved in this work? Summing up the problem that seemed to be at the root of it all, an audience member asked Oram whether there needed to be "a more human approach to electronic music." She responded: "In my little illustrations tonight, I've tried to make them perhaps more human, and show that one can produce all sorts of feelings in this, by this method. Perhaps if we will be able to get more humanity. Perhaps if we drop the slide rule, we may get more human feelings. That's a personal opinion of mine, perhaps not what is thought in Italy and Germany."[50] While endorsing the musicality of all the pieces she presented that evening, Oram also suggested that serial techniques amounted to composition by "slide rule," governed by external mathematical measures rather than human feeling, perception, and comprehension. It was, moreover, with former axis powers Italy and Germany that she associated such an inhuman method, in contrast to her own belief in the need for human-centered approaches.

The difference between "slide rule" and human-based approaches was not only manifest in serial techniques, however. One also needed to consider the differential between the instrumentally measurable and the humanly perceivable in sound. As Oram observed with regard to the frequency range of human hearing measured in hertz (Hz): "You might think we could have 15,000 notes in our scale system. But this of course is not true because the ear just does not comprehend any difference between 400 and 401." Such discrepancy between the theoretical possibilities according to mathematics and physics and the scope of possibility according to physiology and psychology was at the heart of Oram's thinking about the need for exploration and experimentation in electronic music. She concluded:

> I would urge more research into finding out not only what scale systems, what pitches we can have, but also what rhythmic figures can we have. Does the human mind rely on 2 beats, 3 beats, 4 beats? Is there something that relates back to that sort of rhythmic pattern inside us? Are we going to always try and relate a rhythm to that? If it doesn't fit in, are we going to discard it, or not? We in 1961 may not appreciate things that will be appreciated in another 100 years. Can we educate our ears to appreciate much smaller intervals? Perhaps to appreciate rhythms which at the moment seem to have no rhythmic basis? That I think is just for the future to say.[51]

Spontaneous applause greeted this remark, indicating agreement from the audience—an optimism that while they might not know how to find meaning in electronic music yet, the new medium opened up the possibility of refined and expanded musical sensibilities, pointing toward a day when their future selves or future generations would be able to perceive and comprehend more than ever before in human history.

A SOUND WAVE INSTRUMENT

The "more research" necessary to bring about such a future was precisely what Oram hoped to do. Thus, as she built a career as electronic composer-for-hire and lecturer in the early 1960s, she also pursued funding to free her from such activities so she could instead devote her time to research. Eleven months prior to her Edinburgh Festival presentation (in October 1960), Oram submitted a proposal to the Calouste Gulbenkian Foundation (CGF) with her research goals and rationale. Again contrasting "music-by-slide rule" and "mathematical formulae" unfavorably with "organization by the human mind," she proposed to assess "the powers of the human ear and mind to comprehend acoustic sensations outside those normally employed in Western Music," to design "electronic circuity to satisfy the requirements" of that assessment, and to apply the results of both to "composition techniques."[52] Through a back-and-forth with the Gulbenkian Foundation, Oram reined in her more all-encompassing program for the development of electronic music and arrived at a specific funding proposal to develop a "sound wave instrument."[53]

Developing some kind of sound wave instrument had been among Oram's original aims when setting up her independent studio in 1959. Announcing her new studio, Oram publicized her goals of "1) synthesizing timbres of a more complex musical nature than those now produced by electronic generators [and] 2) providing the composer with a direct link between pen and sound."[54] One of Oram's sources of inspiration for this vision for electronic music was *Music for All of Us* (1943), a book by the celebrity conductor and broadcast media innovator Leopold Stokowski.[55] From Stokowski, Oram drew the prediction for a future day, "when the composer, instead of writing notes into a score, will compose directly into *tone*—just as a painter composes his picture directly into color"—and "will have at his disposal *all* the timbres that are possible in Nature."[56]

Another idea Stokowski discussed in *Music for All of Us* was the centrality of the hand to musical expression. This centrality emerged from the contrast between hand and machine put at issue by the rise of industrial manufacturing and the experience of living in a world increasingly populated by mass-produced objects. As he explained, "The fact that everything in music is hand-made is of basic importance. It is natural for us to be influenced by the machine and machine-made things. Most machine-made things, like automobile parts, are exactly standardized. This standardization is essential to machines, but fatal to music."[57] The hand/machine dichotomy was (and remains) a recurring theme for critics of the transformations to labor and material culture wrought by automation. Another musical example can be found, for instance, in George Gershwin's 1930 essay "The Composer in the Machine Age," which concluded: "The composer has to do every bit of his work himself. Hand work can never be replaced in the composition of music. If music ever became machine-made in that sense, it would cease to be an art."[58]

Though Oram did not mention Stokowski's theory of the hand directly, her writings reflect a shared configuration of hand, imagination, art, and human control set in opposition to machines and standardization. Describing how a composer would specify various musical parameters graphically to have them rendered in sound by the instrument, she paused to reflect: "Here I am verging on electronic computer technique and the robot machine, which may seem contrasting to all artistic requirements. But I hasten to put your fears at rest at this point. The composer here is deliberately pre-setting a chain of events which will from then on occur automatically. In presetting it he has all the elements under his control." Oram suggested her envisioned method was more like composing for orchestral instruments, wherein "much of this presetting is already done for" the composer by virtue of the instruments' materiality (the variation of timbre with register in the clarinet or bassoon, for instance, or with the different strings on a violin). She concluded: "Electronic computer technique need not mean composition by dehumanized mathematical methods but can mean that the composer has everything under his control and can freely exercise his imagination."[59]

TOWARD HUMANIZING THE MACHINE

Oram began her Gulbenkian-funded work on developing the sound wave instrument in 1962 but made slow progress until 1965. During these years, her reports to the Gulbenkian Foundation largely concerned how she was laying the groundwork for using the sound wave instrument once completed to research the effects of sound on health. The outlook changed when she hired a new engineer, Graham Wrench. In February 1965, Oram sent a report to the CGF, updating them not only on the hiring of Wrench and his technical plans for realizing the sound wave instrument but on their reoriented priorities for its use:

> As you will know from my previous report, I have been keen to explore the possibilities of music therapy with the Oramics equipment. However, after much discussion with Mr. Wrench, it would appear that, first of all, instead of exploring the medical application, there is a huge void to be filled in the artistic sphere—the study of the emotional acceptance or rejection of auditory sensations. This demands that the equipment be developed in such a form that the moods and emotions which the composer desires to express will be truthfully reproduced in the final product without the intrusion of any distortions, or any "coldness," caused by mechanized control. Every nuance, every subtlety of phrasing, every tone gradation or pitch inflection must be possible just by a change in the written form.[60]

There are subtle but significant changes in this explanation of the contribution of Oram's sound wave instrument to electronic music as an art form, compared to her earlier funding requests. One change is the target of criticism. Previously, "mathematical methods" and automation were the figures of dehumanization—of removing the human from music making—to which Oram contrasted her approach. Here, the threat to musical art appears in the form of electronic equipment that would contribute its own qualities—"distortions" or "coldness"—rather than faithfully render the composer's every musical nuance and emotion. The shift of focus to the technology as either source of inhumanity or (when properly designed) transmitter of humanity set the stage for reframing Oram's research; rather than dispense with "the slide rule" in favor of the mind, her work would help reconcile the seemingly antithetical spheres of science and art:

> We wish to show that science can assist the Arts without inducing, cold, calculating, lifeless, mechanical results. We also wish to show that machines, however complex they may be, have not of their own accord produced—and we believe they never will be able to produce—a work of Art which has that indefinable difference, the stroke of genius. This is a product which stems from an inner inspiration that we trust no machine will ever fathom. However, we feel that machines can provide new ways in which that inner inspiration can express itself; they can help to give a greater understanding of the medium and can aid the study, in minute detail, of the stimuli and the sensations produced—for it is our belief that by a greater knowledge of the result we may better appreciate and respect the tremendous unknown—inspiration.[61]

Rather than contrasting her "instrument" with "the robot machine," Oram now argued for the viability of machines for artistic expression. Updating Quantz's contention that a musical machine might astonish but would never move you, Oram drew the line at fathoming inspiration or creating genius works of art autonomously. Through its precision in "stimuli" and "sensations," however, a machine could be an instrument of science and art at once.

During the brief period (under two years) that Wrench worked with Oram, he succeeded in building a functioning instrument that converted drawn information into sound, with graphic control of timbre, pitch, duration, vibrato, and reverberation.[62] In June 1966, the sound wave instrument—which she now called the Oramics equipment—was finally functional. Rather than publicize this milestone, however, Oram reported to the Gulbenkian Foundation that there was still "much work to be done in learning [the system's] 'language.'"[63]

In 1968, an event billed as the First London Concert of Electronic Music by British Composers became the occasion for the first public outing for Oram's drawn sound. Held at Queen Elizabeth Hall on January 15, the concert included Oram's "Contrasts Essconic," a composition for piano and electronic tape. A contraction of Electronic Studio Supply Company, Essconic was the name of a company Oram had formed for her work on drawn sound. The "contrasts" of the piece were in sound sources, qualities, and mood. As she explained in a program note, the composition used "sources ranging from simple and complex tone generators, through 'musique concrete' resources to 'Oramics' (graphic sound, the invention of Daphne Oram)." The piece thus accomplished in one composition the kind of survey of electronic music techniques and advocacy for their ability to "produce all sorts of feelings" that she had presented at the Edinburgh Festival, adding to that survey her new graphic sound. The most likely passages in which to hear graphic sound occur at 3:56–4:50 and 5:40–6:43, where the wandering, warbling ethereal tones and reverberation are reminiscent of both the A section of "Four Aspects" and her later piece, *Brociliande*, which she identified as being made with Oramics and "everything written by hand" (audio 19).[64]

Oram's piece was programmed between two by Peter Zinovieff, which contrasted sharply in their emphasis on computer automation and overshadowed the rest of the program in media attention. Zinovieff's "December Quartet," according to the concert program notes, was "composed and realized entirely by computer." For "Partita for Unattended Computer," "a large amount of freedom has been given to the computer" so it could deliver a "live performance." The BBC radio program *World at One* promoted this work as the "first concert performance ever by a computer." Speaking on the program, Zinovieff explained: "This is the first time the computer's actually produced the sounds itself, or controlled the—not only composed but also controlled the apparatus that produces the sound."[65] It was "Partita" that also featured on the BBC One television program *Tomorrow's World* episode on computers and electronic music, which aired a couple months after the concert.[66] Focused on Zinovieff, the episode depicted his compositional process as writing by hand a score in "a new form of notation, new because after the work has been composed it has to be translated into the language of computers." It is then a "secretary" who "types the composition out" in a set of numbers, which are rendered as a punched tape that is fed by the secretary into the computer's "memory bank" (an unnamed woman is shown performing these tasks). The *Tomorrow's World* episode also included footage of the "Partita's" performance at the Queen Elizabeth Hall concert. To begin the performance, Zinovieff and a woman (identified elsewhere as Delia Derbyshire, whose composition *Potpourri* was also on the concert) checked "that the composition is correctly loaded into the computer" and started it going. To do this, Zinovieff stood, appearing to adjust knobs amid the computer's lights and cables, while Derbyshire sat, like the secretary in *Tomorrow's World*, at a computer keyboard. Zinovieff and Derbyshire then walked off the stage in opposite directions and the computer continued alone, sounding (after a more varied introduction) a monophonic sequence of tones of equal loudness, duration, and sine-wave-like timbre, leaping unpredictably according to Zinovieff's predefined parameters to random effect. The computer performed, in other words, what became the musical topos of computers "thinking" in film and television. In a shot of the listening audience, Oram is visible, but no mention is made of her or any other composer's work.

Later the same year, a special exhibition at the Institute of Contemporary Arts (ICA) entitled *Cybernetic Serendipity* presented another opportunity for electronic composers to showcase their work. In a July 1968 letter to composer David Lilburn in New Zealand, however, Oram wrote that "everything [was] to be strictly Aleotoric [*sic*], so that's ruled me out." She complained: "At present here, one only has to mention the word computer for everyone to swoon away in wonderment, leaving behind them all their critical faculties. The computer can then produce 'music by the yard' and get away with any rubbish. It is depressing for me, having spent some years devising 'computer like equipment' as an 'extension to the composer's arm,' responding only to the minute instruction of the composer."[67] The letter conveys Oram's sense of being at odds with, and excluded from, the direction of computer music then receiving institutional support and being represented in the media. While Oram had no work in the *Cybernetic Serendipity* exhibition or on its accompanying album, however, she was not completely absent, for there were also two issues of the ICA's monthly magazine published during the exhibition and devoted to its theme. Oram contributed an essay to the second of these. The essay included several photographs of Oramics and the drawings it would turn into sound, including one featuring Oram with pen in hand, inking a wavy line, under the caption "Daphne Oram writing in analogue programme" (fig. 3.3).

Oram used her *Cybernetic Serendipity* essay to make the case for Oramics in contrast to the current direction of computer music. Although Oramics could be considered a kind of computer, she noted, it primarily processed analogue rather than digital information. And unlike the practice in computer music, Oramics required the composer to specify "exactly what he wants." Implicating Zinovieff's approach, she wrote, "For those who would rather get the computer to do it all for them, there is always the random number table and the digital computer—they can then sit back and have music by the mile! But not for me!" Oram grounded her approach in a link between "freehand" drawing and "nuances and subtleties which make the music more alive and less clinical than Electronic Music."[68] She concluded her essay by quoting a critique of serialism in electronic music by composer Tristram Cary, which offered another image of connection between composerly thought and hand and of aversion to computer con-

FIGURE 3.3. Photograph that appeared with the caption "Daphne Oram writing in analogue programme" in her article on Oramics in the *Magazine of the Institute of Contemporary Arts*, second *Cybernetic Serendipity* issue (September 1968, p. 11). Daphne Oram Archive ORAM/7/9/013 ("Daphne Oram drawing timbres on the Oramics machine," photography by Fred Wood) © Daphne Oram Trust. Image provided by Special Collections & Archives, Goldsmiths, University of London, with permission from the Daphne Oram Trust.

trol: "O electrons... keep your appointed places—under my thumb."[69] In light of the Queen Elizabeth Hall concert and its media coverage, the Oramics machine presented not only alternative technical solutions and aesthetic values for electronic music but also an alternative to the gendered division of labor on display in Zinovieff's work, where a male genius composer conveyed his thoughts by "analog" handwriting and knobs while a female secretary translated his ideas to machine via the digital operations of her fingers at a keyboard.

At the bottom of the *ICA Magazine* contents page for the issue that included Oram's article, Oram likely noticed a block of text under the heading "MAN-MACHINE SYMBIOSIS": "Humanizing machines and mechanizing humans are cross trends that are sure to occur in the future, but the extent to which man and machine will be united is uncertain."[70] Next to the text, a person wearing a helmet stands poised with hands on

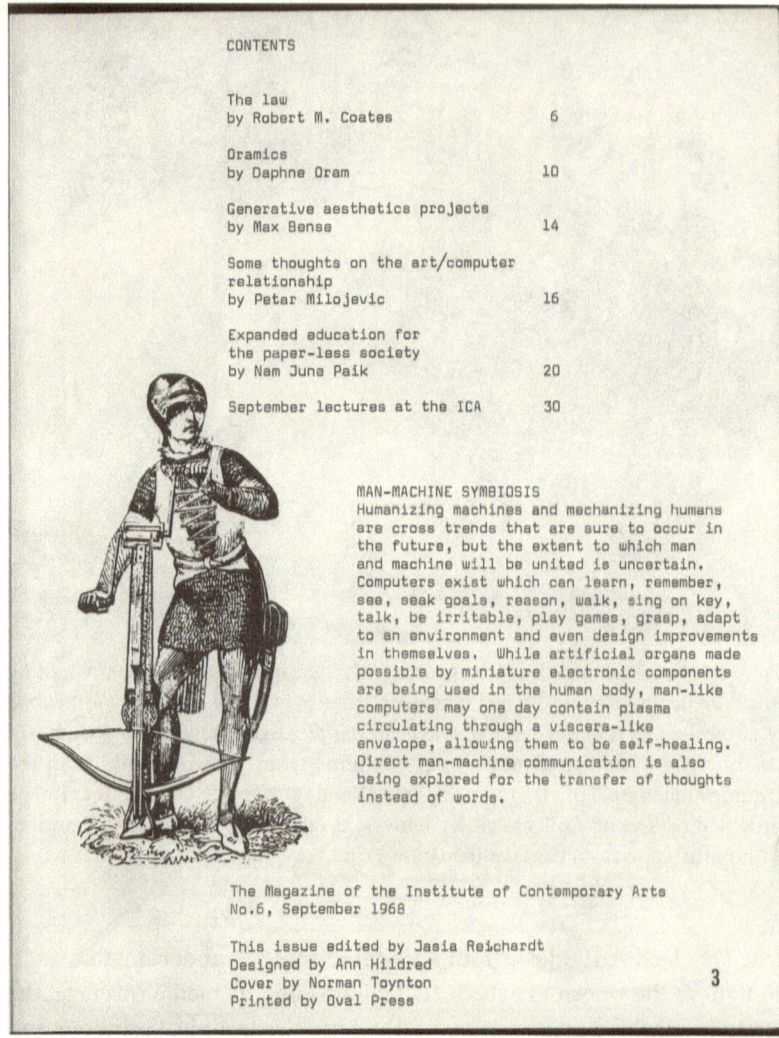

FIGURE 3.4. Contents page for the second *Cybernetic Serendipity* issue of the *Magazine of the Institute of Contemporary Arts* (September 1968). Image courtesy of Princeton University Library.

the handles of a crank, one foot inserted into a loop holding down the opposite end of the device (fig. 3.4). This image reproduced an engraving from the 1880s, which in turn looked to the past to depict a medieval archer with crossbow. "Humanizing machines" and "mechanizing humans," this image suggests, are cross trends that were not only "sure to occur in the future" but also have been happening for hundreds of years. Soon after this exhibit, *humanize the machine* entered Oram's vocabulary as a driving goal of her Oramics work, and she began to think through how *the human* might be explained by electronic sound.

"ARE WE HUMAN BEINGS EACH A VAST NUMBER OF TUNED CIRCUITS?"

Cybernetic Serendipity sparked the formation of the Computer Arts Society (CAS) in 1969, and Oram became a founding member of the society. While one early member noted that using the computer as a random number or sequence generator seemed to be "enshrined in the CAS," the bulletins also reflect a more philosophically and aesthetically diverse artistic community, united by a conviction that "computers can lead to more freedom for artists, not less."[71] The second issue (May 1969) reproduced image and text from a poster by the British Psychocybernetic Organisation that portrayed the brain as a collection of resisters and capacitors and declared, "Your mind is a computer—and right now you're only using 1/10th of its capacity. Hear how you can use your potential to the full through PSYCHO-CYBERNETICS." The bulletin editorialized: "Text and image typify an attitude to man and machine that needs to be opposed" (fig. 3.5). The evident target of their opposition was the reductive image of the human, the computer-brain as image of efficiency and optimization rather than creative possibility. As announced in a brochure on the society's formation, the CAS held that although it would be impossible to "return to the days of hand made goods," it was possible "with computers [to] reverse the tendency for mechanization to mean standardization, greyness, uniformity and de-humanisation."[72]

There is reason to think Oram would have interpreted this image differently, however—not as typifying an "attitude to man and machine that needs to be opposed" but rather as exemplifying how humans' physical makeup made possible greater capacities for perception and consciousness than generally in use and, further, how electronic equipment could model and help explain *the human*. Looking with Oram, we might see in the image an embodied mind, its paradigm not a digital computer but a musical instrument, its primary medium vibration. Such were the ideas Oram developed in her book, *An Individual Note*.

In December 1970, an editor at Galliard Press reached out to Oram about writing a book on electronic music. Oram quickly accepted the invitation and from the start approached the project with ambitions significantly beyond music aesthetic and technical matters. As she wrote to the editor, she envisioned offering not a textbook or introduction but "an attempt to show the potentialities, the excitements of the new medium," so

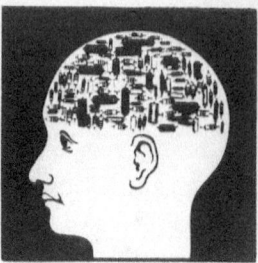

FIGURE 3.5. Front page of the second issue of PAGE, the Bulletin of the Computer Arts Society (May 1969). Daphne Oram Archive ORAM/5/4. Computer Arts Society, BCS. © Daphne Oram Trust. Image provided by Special Collections & Archives, Goldsmiths, University of London, with permission from the Daphne Oram Trust.

that readers would "gain a broader musical outlook, a richer experience in sound [. . .] maybe even gain a hint of some of the basic wave phenomena which seem to permeate everything we know (from the atom to the galaxy, from breathing to expiring!)."[73] As Oram researched and wrote, these ambitions took shape as a systematic inquiry into the human through analogies with sound and electronics. As she summed up on page 52 of the resulting book: "We have been looking at oscillators, ring modulators, filters and formants and musing upon the possibility of their functions being an analogy of what happens within the human brain and body."[74]

Analogy, Oram observed, was "a process much used in the past." In the present day, if art and science were to be brought together, truly assisting each other rather than just transferring scientific materials to artists, "both scientist and artist need a new range of metaphor, verging on mythology: a new set of analogies."[75] Oram's implication that analogy was foreign to scientific thought echoes the suspicion Diderot voiced around his clock bells analogy. However, at the time she was writing, analogy had acquired a new centrality and epistemological respect in the interdisciplinary science of cybernetics.[76] CAS bulletins were a source of information about cybernetics for Oram, providing not only articles by members of the society but also lists of relevant books from which Oram added to her reading lists. One such title was Arthur Siegel and Jay Wolf's *Man–Machine Simulation Models: Psycho-social and Performance Interactions* (London 1969). If Oram succeeded in getting her hands on this book, she could have read in it not only about their computer models for man-machine systems, but also that "it can be claimed with some validity that the story of man's progress in science and technology is actually the story of success in the use of analogy."[77] Siegel and Wolf, in turn, drew this quote from the published outcome of a symposium on computer simulation in engineering and life sciences sponsored by the US Air Force Office of Scientific Research and Westinghouse Defense and Space Center, exemplifying the shared epistemological investment in analogy in cybernetics research.[78] Noting the importance of human-machine analogies to the field, Slava Gerovitch emphasizes their self-legitimizing circularity, arguing that "cybernetics did not merely describe computers metaphorically as brains; the brain itself was conceptualized in logical and engineering terms, and these concepts then returned to computing, serving as a basis for the impressive 'discoveries' of man-machine analogies."[79]

Besides cybernetic discourses, there were other sources informing Oram's use of analogy. At the time of her invitation from Galliard, Oram was also already formulating ideas for a lecture she called "Waves and Wonder," to be delivered to the Churches Fellowship for Psychical and Spiritual Studies. Founded in 1953, this community was dedicated to studying paranormal phenomena scientifically and for medical applications and integrating these phenomena with Christian beliefs.[80] Oram's lecture concerned "music, electronics & metaphysics," and her notes show her using many of the concepts she would cover in her book, including the terms *ELEC* for a material flow starting from an electrical spark or tension and then dissipating and *CELE* (*ELEC* backward) for a spiritual flow by which something beyond the material emerged into being.[81] In her lecture notes, Oram asks, "Are WE HUMAN BEINGS each A VAST NUMBER of TUNED CIRCUITS?"[82] And further, "Can we use our knowledge of sound, of music, of psychics [NB: not physics], of electronics, to create analogies,

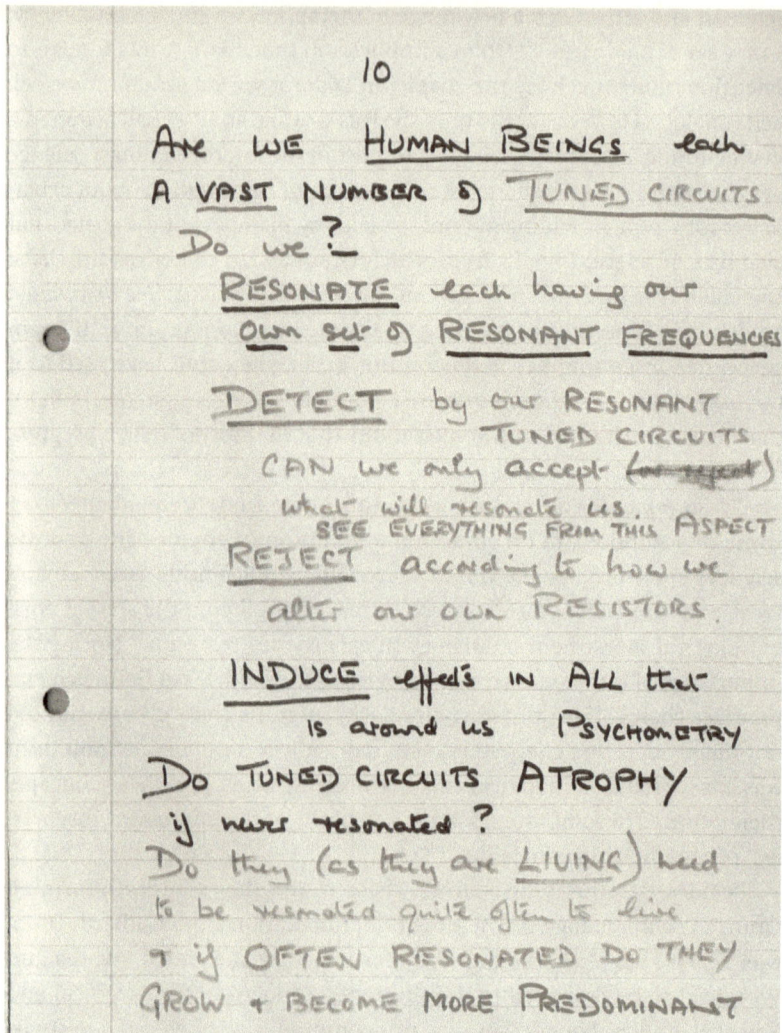

FIGURE 3.6. Notes for the "Waves and Wonder" lecture Oram delivered to the Churches Fellowship for Psychical and Spiritual Studies, April 1971. Daphne Oram Archive ORAM/4/7/001. © Daphne Oram Trust. Image provided by Special Collections & Archives, Goldsmiths, University of London, with permission from the Daphne Oram Trust.

metaphors, parables which will assist us? Indeed I think we can do more—we can use these very materials to Work out the Ascendancy of the celestial" (fig. 3.6).

These lecture notes show Oram's analogies participating in a longer tradition, wherein electronic technologies provide models for explaining how communication with the spirit world and other supersensory phenomena are possible. Jeffrey Sconce and Laura Otis among others have investigated this tradition, demonstrating the codevelopment of Spiritualism and telegraphy in the nineteenth century and the contributions of psychical research to the conceptual imbrication of individual nervous systems with global telecommunications systems.[83] Richard Noakes has shown how the spiritualist séance, as exemplary site of spiritual communication, generated debates over whether human mediums or scientific instruments provided more sensitive and reliable detectors of spiritual presences, with defenders of Spiritualism using "analogies between séance bodies and scientific instruments" to support the reality of manifestations from beyond.[84]

For her spiritual-scientific conception of the human as "a vast number of tuned circuits," Oram specifically drew inspiration from the work of Shafica Karagulla. When Galliard invited Oram to write a book on electronic music, Oram had already read Karagulla's *Breakthrough to Creativity* and reached out to the Beverly Hills–based doctor to express her agreement with her theories and interest in working together. By training a psychiatrist and psychiatric researcher, Karagulla was inspired after reading about precognitive and telepathic abilities to set up her own research center, the Higher Sense Perception Research Foundation.[85] In *Breakthrough to Creativity*, Karagulla positioned her research on higher sense perception in the context of a broader transition in human consciousness, from "a world of static solid forms into a world of dynamic energy patterns," describing the latter as "a universe that is nothing but frequency." A bacteria or virus, for example, was "in the last analysis, a frequency," and drugs used against them were "also frequencies," suggesting that when bacteria develop resistance to medications, it may be due to an ability "to change their frequency pattern." According to Karagulla, the new consciousness of the world as energy patterns and frequencies prepared "man" to develop a "'sense of frequency,' which could give him a more direct experience of his real environment."[86] Karagulla thus shared the sense of potential Oram had expressed at the conclusion of her Edinburgh lecture for humans to expand or transform; and with the concept of frequency, she explained that potential in terms that were at once specifically musical and universal.

Oram's knowledge of musical electronics, however, led her beyond Karagulla's comparatively simplistic and disembodied notions of frequency. To explain how something in the world reaches consciousness, for example, Oram drew an analogy to ring modulation. A ring modulator, Oram explained, is a device that accepts two waveforms and intermodulates them, such that it outputs the sum and difference frequencies of the two waveforms. Input signals of 440 Hz and 600 Hz, for instance, would output 1040 Hz and 160 Hz. Oram suggested that consciousness receives not the input signals but the output signals of intermodulation. As she described the biological process in electronic terms: "Acceptor circuits [within us] do not receive direct signals from outside, but they receive signals which are resultants... the outcome of the internal personal signals intermodulating with the incoming external signals" (ellipsis in original).[87] If this is this case, she asked, "however much you shift your own wavepattern over the frequency spectrum, will you ever actually *see* the object, with which your own wavepattern is intermodulating? Is this an analogy of life? Do we ever perceive reality?"[88] Rather than suggesting a "sense of frequency" would give one a "more direct experience of his real environment" as did Karagulla, Oram argued that to gain even "a glimmer of the reality of the outer world" would require "complete formant control of our own wavepattern" and developing the ability to modify one's "various regions of resonance" such that, through the shifting intermodulation with external signals, one would be able to deduce reality.[89]

VIOLIN AS EXEMPLARY HUMANIZED MACHINE

In the publicity that surrounded the publication of *An Individual Note*, Oram did not discuss her analogies between humans and electronics, perhaps considering them too esoteric for general audiences. Instead, she focused on how she was "humanizing the machine." On the BBC radio program *World at One*, Oram spoke with Nicholas Woolley about—as he introduced the segment—how she "developed a method of translating visual wave patterns into musical sound, in a way that retains the human element." Demonstrating the process with a melody used in *Bird of Parallax* (composed for a ballet), Oram explained that she was "wanting to make some sort of trumpety sounds, but not exactly a trumpet," and so she "drew a wave pattern, and I gave that to the machine, and then I went and put some black dots on. Now, this just gives me a number of pitches one after the other. Let's hear those." There followed a series of notes with straight tone and even volume, after which Oram commented, "very dull. Now we want to put vibrato, and we want to put reverberation, and we want to rephrase it and make it much more musical." The same series of notes then repeated, this time with reverberation throughout and expressive additions such as vibrato on the penultimate note.[90]

To my ear, what Oram describes as "very dull" sounds like a rather chaotic, random series of notes, which acquire a sense of direction and purpose in her "more musical" version. Most decisive in effecting this transformation is a long-held pitch that Oram "rephrased" to be a short note repeated and accelerating into the next note. This moment confounds what might seem like a traditional division between composition and expressive performance. The idea of being "very dull" also meant more to Oram than simply lacking expression. In lectures, she used "DULL!" to describe sounds with "very little information," like a sine wave. "Just as dull," however, was a sound with too much information, like white noise with its random frequencies over the whole spectrum. Sine wave and white noise represented "two somewhat hazy boundaries" for Oram, within which to chart the composer's territory of sound and discover "the information needed in a note. What is necessary? What comprehended?"[91] One necessity Oram advanced in *Individual Note* was variation in any musical note longer than about a second, since it is not "human" to "maintain anything in a steady state for long." Like limbs, eyes, voice, and brain, musical notes need to "vary subtly throughout their duration," or else they produce fatigue.[92]

Following the musical demonstration, Woolley teed up Oram to address the concerns listeners might have about what new electronic machines were doing to music: "I think a lot of people are rather worried by the complication, apparent complication of electronics. But is your equipment really *just* an instrument in the same way that a violin is an instrument?" Noting that the violin had had 600 years to "evolve," Oram explained that her equipment similarly needed "to evolve slowly and not be pushed." Intoning *human* with a special slowness and inflection, she added: "I've got to find out how it's possible to make it human, in a way. I mean the fiddle to me is a marvelously human instrument, it's almost an extension of the human being. And if I can do this, machines can I think be humanized, not only music machines but machines in other fields."[93]

Oram's preparatory notes for this interview show her planning the comparison between the Oramics equipment and the violin, which she described as a "good example of machine being humanized." She also cited a 1922 edition of *Grove's Dictionary of Music* for the observation that the violin is "a machine just as a watch, a gun, or a plough is a machine. But it's evolved," and by becoming "more supple and subtle as a means of conveying musical thought," it has become "part of the human being who is playing it."[94] Though not cited in her radio interview notes, Oram likely also drew inspiration from the March 1972 CAS bulletin, which included an article that similarly used the violin to rethink "the machine." The article, "Reflections on Art," was authored by John Whitney, a California-based artist developing computer animation techniques, who also had training in musical composition. Oram marked (by way of a vertical line down the margin in her copy) the passages underlined in the following:

> Observing strictly the dictionary definitions of both "instrument" and "machine" it is not false or incorrect to call the piano and all the instruments of the symphony orchestra: "machines." How well, or just how, the professional musician plays upon those "machines," we might call man-machine communication, for the sake of a fresh look at an old subject. To do so might establish a new, if unorthodox, set of criteria for the general subject of man-machine communication as it applies to the computer sciences.
>
> Just how well are we doing with man-machine communication compared with man and violin? The answer has to be: poorly! In fact it is doubtful if there could be a more intimate, more sensitive, more perfectly interactive relation between man and machine than that which obtains between master and piano or violin.
>
> I am drawn to propose new sensitivity criteria such as this in order to suggest that if computer systems are to be used at all for art purposes they may have to match criteria already quite old and established in the art of machines, or musical instrumentation field. Therefore, all aspects imaginable need fresh consideration; control, response, feedback, sensitivity, rapidity, etc.[95]

By including musical instruments in the category of machine, the violin became the quintessentially "humanized machine," a model to be taken up in computer science for its "perfectly interactive relation between man and machine." Whitney's essay agreed with the arguments Oram had made for nuanced control of computer-like equipment in her *Cybernetic Serendipity* essay and *An Individual Note* and offered a new way both to express the idea and contextualize her invention historically. Conceptualized as a machine, the violin became a historical precedent for Oramics as a machine in an early stage of technological evolution toward becoming fully "humanized," an "extension of the human being."

"THE *AS IF* SPIRIT"

Cypess and Steven Kemper have examined historical recurrences of what they term the "anthropomorphic analogy" in musical culture, embracing with this term comparisons of musical instruments to the human body and mind, humanoid robotic instruments, and "cyborg" musical systems in which performers wear sensor-based interfaces. Comparing instances in the seventeenth century and today, they show how anthropomorphic analogies express "a desire to humanize" technologies, how they problematize the relationship between humans and technology, and how they prompt reconsideration of "the nature of the human body itself."[96] Diderot and Oram both fit this picture and add to it. Their analogies show a history not just of speculation about sentience reflecting contemporary machines, or of efforts to humanize technologies, but something more like local cross-trends: humanizing harpsichords and harpsichordizing humans; humanizing Oramics and Oramicizing humans. Called on to explain the otherwise inexplicable, the analogical relation creates this back-and-forth movement.

At the outset of this chapter, we observed how a convergence on one analogy—between brain and computer—facilitated a slippage from *as if* to *is*. A benefit of more analogies is the ability of each to refract what the others clarify and obscure, to hold in view what likenesses are explanatory and what they carry along by implication or extended inference. Bringing Diderot's harpsichord and Oram's sound-wave instrument into the history of machine-based speculation about the brain offers to reactivate the *as if* of digital computers and information processing, to resituate the field of artificial intelligence as one local cross-trend rather than the successful culmination of that history. Their musical analogies contest the category of machine, and materialize a sounding human.

FOUR

Personifications

Piano Death & Life

What is happening in this photograph (fig. 4.1)? On the front page of the Monday, July 30, 2012, issue of the *New York Times*, the photo appeared with the caption: "Bryan O'Mara tossing out a piano in Southampton, Pa." But the accompanying article by reporter Daniel J. Wakin told a different story, one in which the pianos were no mere recipients of human actions but rather authors of their own doings: "The Knabe baby grand did a cartwheel and landed on its back, legs poking into the air. A Lester upright thudded onto its side with a final groan of strings, a death-rattling chord.... The site, a trash-transfer station in this town 20 miles north of Philadelphia, is just one place where pianos go to die."[1] In Wakin's anthropomorphic language, the piano is not the passive object but is the active subject of the picture. On its deathbed, the piano comes to life.

Wakin was not alone in perceiving the piano as more animate than inanimate, more person than thing. Readers wrote in with tributes to their own instruments: "I thought of that piano as my mom's beloved friend," wrote one. "A wonderful family member for 50 years in our home," another said of a Baldwin medium grand. "A person does feel a bit silly treating a piano like a pet," reflected another on her decision to withhold her instrument from a new home where it would likely be ill-treated, "but goodness, I can't help it!"

We have seen the piano declared "the most appalling contrivance of levers and wires this side of the steam engine," when Barzun used music's long "machine-ridden" history as an argument for giving electronic music a chance. Whereas casting the piano as a machine served to pave the way for a new generation of music technology, regarding the piano as a beloved friend, family member, or pet has an opposite effect, arresting the ability to replace one instrument with another. While the resemblance of pianos to persons might be wielded just as strategically as their machine-like nature, however, there is an unbidden quality to the responses to the

FIGURE 4.1. Photograph on the front page of the *New York Times*, July 30, 2012. © Jessica Kourkounis. Reproduced by permission of the photographer.

piano as if it were a living being and close relation. "I can't help it," as one commenter remarked, despite feeling silly about putting a piano on the same plane of being as a pet.

In 2000, a team of psychology researchers reported that musical instruments constitute a "crucial exception to the living/nonliving dichotomy" observed in category-specific agnosia. People with category-specific agnosia have difficulty naming certain categories of things as a result of brain damage; depending on the damaged area, the difficulty applies mainly to naming either living things (such as animals, fruits, vegetables) or nonliving things (such as tools, furniture, clothing). In patients with an impaired ability to name living things, however, the researchers found a similarly impaired ability to name musical instruments: "Artifact naming was flawless whereas musical instrument and biological object naming were both severely impaired." The researchers concluded that understanding of category-specific agnosia was "limited by adhering to the spurious living/nonliving distinction."[2] Their finding also opens up social and cultural questions; by what processes would musical instruments come to be categorized as nonliving yet experienced as living?

Of course, not everyone reacted to scenes of piano dumping with shock and dismay. Some commenters on Wakin's story took a more dispassionate view, pointing out that "pianos are consumer goods... [which] will pretty much inevitably become trash someday," that many of the dumped pianos belong "in a landfill replaced by a better instrument," or (in an echo of Barzun's argument) that "technology marches on"—"the harpsichord was replaced by the piano, the piano is being replaced by electronic keyboards... it is called progress." These reactions, premised on the inevitably of piano demise at both the individual level (objects wear out) and the species level (technologies become obsolete), reveal the contingency of more anthropomorphic perceptions of piano being. As Sconce has argued with regard to attributions of animacy and even sentience to electronic media technologies, such notions circulate "not as timeless expressions of some undying electronic superstition, but as a permeable language in which to express a culture's changing social relationship to a historical sequence of technologies."[3] With regard to musical instruments, Eliot Bates observes that "some instruments seem to possess a golem-like autonomy," while others "seem (at least as far as we know) to have little agency."[4] Together, these observations suggest some meeting in the middle of material affordances for and enculturated perceptions of musical instruments as living.

By bringing normally unseen scenes of piano dumping into public view—by putting large-scale piano *death* on display—Wakin's story and others like it have provided a rare platform for collective reflection on piano *life*. In the late twentieth century, there was a ring of magical thinking to observations like Abbate's that a musical instrument is "an object given life as long as a master plays it."[5] Pianos at the dump, however, call attention to instrument "life" that persists or arises outside the parameters of performance. They manifest life apart from any human "master," recasting such dependency not as a magical transfer of animacy but as symptomatic of a limited anthropocentricism.

Anthropomorphism offers one way to think about this phenomenon. As Festa has written, anthropomorphism "replicat[es] human likenesses everywhere." Rather than being a sign of human mastery, however, it is more a response to incomprehension, a recourse to the self in the face of failure to grasp the world by any other means. Jane Bennett, on the other hand, emphasizes the potential for anthropomorphism to register a "sense of a strange and incomplete commonality" with nonhuman beings, which may have the effect of inducing one "to treat nonhumans—animals, plants, earth, even artifacts and commodities—more carefully, more strategically, more ecologically."[6] Particularly in a world where only humans are believed to have agency, Bennett argues that "we need to cultivate a bit of anthropomorphism" as a counter to anthropocentricism and as a conceptual tool for recognizing "echoes" of human agency in nonhuman beings.[7]

There is a problem with anthropomorphism, however: it requires—or produces—a sureness about what is properly human, such that those human qualities can be projected onto beings to which they do not properly belong. This is the dynamic Barbara Johnson finds in her comparative analysis of lyric poetry and legal texts, which leads her to draw a meaningful distinction between anthropomorphism and a kindred trope: personification. Between the grammatical "first person" possessing subjectivity and the "constitutional person" bearing rights, it is clear that a person need not resemble a human. Thus, anthropomorphism depends "on the givenness of the essence of the human," while personification—by putting human and nonhuman persons on equal footing legally, figuratively, or otherwise—"threatens to make the lack of certainty about what humanness is come to consciousness."[8]

Personification thus provides an alternative to both anthropomorphic conceptions of musical instruments (where human forms or qualities are projected onto them) and prosthetic conceptions (which celebrate performers while reducing instruments to extensions of their being). Reactions to scenes of piano destruction extend a long history of personifying keyboard instruments—of regarding them as having their own souls and figuring them in interpersonal relationships. They also reflect relationships to piano materiality that have shifted with changes in the piano industry and have to do with the specific context of the early twenty-first century, when the survival of traditional pianos appears to be threatened by their digital substitutes.

PIANO MORTALITY

Destruction of musical instruments has become a familiar part of experimental music, thanks precisely to the capacity of musical instrument destruction to elicit strong reactions. After cringing at a performance in which a gamelan was disassembled and its pots filled with water, philosopher Stephen Davies reflected on musical works in which instruments are abused: "We could not be made uneasy or shocked by such behavior unless we were disposed to think there is something wrong about damaging or destroying musical instruments. The artists concerned deliberately set out to exploit that attitude of concern, either to horrify the audience members for the sake of appearing outrageous or to jolt them into noticing an art-political point."[9] Making similar observations about instrument-abusive works by Fluxus artists like Nam June Paik, Philip Auslander identifies "violence against violins and pianos as a specifically cultural ritual, the object of which is to desecrate an aesthetic order—that of high art—by smashing its sacred artifacts in its own sacred spaces."[10]

It is one thing to witness such smashing of artifacts on the elevated plane of performance art, however, where it is ultimately governed by authorial intent and in the service of aesthetic and political effect. It is quite another to see instruments destroyed in the uncivilized act of trash disposal, their fate determined by market forces and practical exigencies. The dump, as Wakin wrote, is where "pianos go to die." In the 2013 opera *The End* by Keiichiro Shibuya, the digital voice and character Hatsune Miku ponders, "Am I going to die? . . . Being mortal makes me almost human."[11] The discarded piano poses a similar question and response. But it is shadowed by an additional question: Is piano death, like its human counterpart, inevitable?

Piano technicians today put the typical life span of a piano at eighty to ninety years. With time and use, especially with changes in humidity and temperature, piano materials deteriorate—wood weakens and cracks, tuning pins come loose, felt and leather wear thin, strings break. When an instrument is considered beyond repair is not just a matter of material deterioration, however, but also of economics. As the house organ of the Piano Technicians Guild puts it: "Not all pianos are *worth the expense* of reconditioning or rebuilding" (emphasis added).[12] For most pianos, the eighty- or ninety-year mark is about when material deterioration has seriously compromised tuning, tone quality, and touch, and the cost of restoring the instrument to playable condition would exceed the cost of replacing it with a new instrument of comparable quality.

Piano haulers report an uptick in piano disposal in the early twenty-first century, which tracks with the life span of the instruments and the fact that sales of new instruments reached their peak in the early twentieth century. Pianos becoming trash is not a new phenomenon, however. In fact, piano-destructive performances stretching back to the 1960s have been predicated on the ready availability of discarded pianos. Such was the case for Lockwood's "Piano Burning" (1968), which unfolds as its title suggests. "Piano Burning" became the first in a series of "scores for piano transplants," all of which involved situating a piano outdoors (hence "transplanted" from its native indoor environment) and awaiting its ultimate destruction (another features a "piano drowning"). Lockwood prefaced these scores with the note, "All pianos used should already be beyond repair." And as Lockwood explained to an interviewer about the genesis of "Piano Burning": "I happened to know that there was at that point a particular garbage dump in Wandsworth, London, which specialized in pianos that people wanted to get rid of. It was a piano graveyard basically, all uprights that people's grandmothers had owned, which were long since defunct and replaced by the telly. So I knew that pianos would be available."[13] Notably, Lockwood places the piano in technological sequence with the television, the newer electronic source of entertainment replacing the older instrument. In this light, Lockwood's piano-destructive works seem not to desecrate an aesthetic order but rather to salvage an already cast-off object and invest its end with artistic significance.

The Baker House Piano Drop—a tradition at the Massachusetts Institute of Technology wherein a piano is dropped from the roof of Baker House dormitory—is similarly a product of readily available discarded pianos. The first Piano Drop took place in 1972, at which time a Baker House tutor, Steve Leighton, collected free or inexpensive pianos advertised in the newspaper and fixed them up in order to supply every floor of the dormitory with a piano. A by-product of Leighton's project was the presence of unsalvageable instruments in the dorm. Student resident Charlie Bruno came up with the idea of putting an unplayable upright to use by dropping it from the roof.[14] (A photo of the original Piano Drop suggests Bruno took inspiration from the falling piano gag featured in Looney Toons cartoons: the word *ACME*—the company from which Looney Toons characters acquired supplies—is stamped on the piano.) Accounts of more recent iterations of the Piano Drop consistently describe the pianos' conditions as "unplayable," "nonfunctioning," or "irreparable."

Videos of Piano Drops now circulate on YouTube, drawing comments from viewers around the world. These comments tend to follow a pattern, with many viewers expressing outrage at the piano drop, often in terms of disgust or disappointment in the people and culture that would allow such wasteful destruction of an instrument that someone would want to play or restore. A handful of others reply repeatedly to these comments to explain that the piano dropped was properly considered trash, being unplayable and beyond economically viable repair. Some initially outraged commenters accept this explanation; others do not, insisting that a piano is never beyond repair, never beyond its musical usefulness.[15]

The feeling that a piano should never die and the sense of surprise that a piano would ever be thrown out with the trash permeate stories about pianos going to the dump. The front-page placement of Wakin's story for the *New York Times*—pianos destined for the dump transcending the arts and culture section—suggests a newsworthiness proportional to the shock value of the phenomenon. A 2016 radio documentary on discarded pianos in Vancouver puts the unforeseen nature of discarded pianos into words: "It turns out that like all living, breathing things, pianos are not immortal."[16]

The idea that pianos should be immortal emerged in the early twentieth century, amid dramatic transformations in the piano industry. Jody Berland has argued for the significance of the rise of player pianos, which promised equivalent access to music without all that tedious practicing. Player piano manufacturers such as Wilcox & White encouraged those who already owned a piano to trade in their instrument for the new kind that everyone (not only those who had learned) could play. In the rivalry between player and traditional pianos, Berland identifies an exemplary instance of obsolescence as theorized by Marshall McLuhan. For McLuhan, a medium's becoming "obsolescent" was not its ending, but a beginning; as Berland explains, "when a medium is displaced by a new medium, it becomes a work of art... its former transparency as a medium disappears behind its newly foregrounded materiality."[17] Thus, with the rise of player pianos, the positioning of pianos shifted: where emphasis had been on "the instrument as a marvelously intricate manufacturing achievement, advertisers now associated the acoustic piano with fantasies of exquisite taste and individual expressiveness."[18]

Berland's suggestion that player pianos prompted a transformation of the piano from a "medium" or "technology" into a "work of art" is compelling but requires amendment on two counts. Firstly, competition with both player pianos and cheaper pianos spurred identification of some pianos as works of art. According to a *Chambers's Journal* article of 1849, the piano existed exclusively as "an heirloom of the wealthy."[19] Within the decade, however, the spectrum of piano prices was widening, and the new availability of cheaper pianos was celebrated for the broader access to music it made possible. In 1855, *Chambers's Journal* described instruments of "fine tone and modest price" available to people of "small means" such as the "needy clerk, the poor teacher, the upper-class mechanic." These modest instruments it declared "the very test and triumph of the pianoforte," comparable to the significance of the "daily press and cheap literature of the nineteenth century" after the "darkness of that time," when a scholar had to transcribe the classics by hand and a parish's one and only Bible was chained to a reading desk in the church. "We should not be sorry to see pianofortes sill more cheaply wrought," the writer added, so that they might be found even more frequently among the poorer classes.[20] English musician and antiquarian Edward F. Rimbault concurred with this view, writing with enthusiasm in his 1860 history of the piano that "men of intellect are beginning to turn their attention to 'cheap' pianos; new and more simple actions are being invented; and the dawn of that day is visible when the 'box of stretched strings,' giving forth sweet sounds, shall be in every man's house, his comfort, his solace, his companion—aye, his *friend!* Let us then look forward to that day."[21]

In order to compete with cheaper instruments, manufacturers sought to educate and persuade consumers about the superior quality of more expensive instruments. Beckwith (an imprint of Sears, Roebuck & Co.) warned about a proliferation of manufacturers of cheap pianos who exploited consumers' ignorance to charge high-quality prices for low-quality products.[22] A 1909 puff piece for Knabe pointed to a "difference between a piano created as a painstaking work of art and most of the instruments that are usually considered to be 'high grade.'" The article countered the immediate appeal of a cheaper instrument with the notion of the piano as an investment that would hold—if not indeed increase—its monetary and artistic value indefinitely: "Those of us who have to consider ways and means are the most concerned to find the piano that will endure, that will never have to be replaced, that will be increasingly through the years to come a delight to the ear, the eye, the touch—*a permanently valuable addition to our homes.*"[23]

Arguments for differences in piano quality required attention to, and a degree of expertise in, the materiality of the instrument. Ultimately, however—and here is the second revision of Berland's argument—rather than newly foregrounding the piano's materiality as McLuhan's theory of "obsolescence" would predict, the more persuasive repositioning of the piano as work of art involved its dematerialization. In the 1920s, Steinway launched an advertising campaign that linked its instruments to the "Immortals of Music"—figures like Richard Wagner, Franz Liszt, Sergei Rachmaninoff, and Arthur Rubinstein—who favored Steinway instruments. Previously, Steinway ads had typically shown images of their instruments in full view. The "Immortals" campaign, by contrast, featured portraits of great musicians at the keyboard with most of the instrument out of frame—when they showed an instrument at all. In some instances, the musician's portrait was replaced by a painting from the "Steinway Collection" of art. Similarly, where advertising copy had previously foregrounded Steinway's patents and manufacturing facilities, such material conditions were now only mentioned to be transcended. As one ad explained, "When you buy a piano you do not buy a thing of wood and steel, of wires and keys—it is music that you buy—the greatest of the arts."[24] Whereas Steinway promotional material previously gave advice on instrument preservation, noting "it is evident that if the piano is to remain in good order for many years, good care must be taken of it," it now simply declared Steinways "the immortal instrument of the Immortals of Music."[25]

What is clear in online reactions to piano dumping and the MIT piano drop is that, for many, the expectation of immortality applies indiscriminately to all pianos, regardless of quality or condition. This stance thrives on a certain distance from the instruments—on the piano as an idea and symbol of music rather than as a material object useful only insofar as it is able to realize specific musical objectives. But dismay at discarded pianos also thrives on a degree of closeness, an attachment to the piano as a family member or friend, a kind of person. And so, part of the shocked reaction to piano dumping stems from surprise that pianos are not immortal. Another part stems from the indignity of the pianos' end—from the affront caused by evident disrespect for the sanctity of piano *life*.

PROSTHESIS OR PERSON

As forms of instrument destruction, piano dumping and musical works that make piano abuse a part of the performance both put at issue a sense of piano *life*. From whence does that life derive? Writing of instrument-destructive works, Davies argues that "we respond to the misuse of musical instruments in respects that are like our reaction to human injury... because we view the musical instrument as extending the musician's body and inner life."[26] Davies draws support for this view from Lydia Goehr, who showed that as part of a nineteenth-century effort by instrumentalists to elevate their performance to the level of singing (which represented true musicality), players "began to speak of their instruments as humanized, as biological, as expressing the inner qualities of human souls.... Instruments were being thought of as immediate extensions of bodies; bodies extensions of souls."[27]

There is ample nineteenth-century evidence for this prosthetic *extension of performer's soul* conception of musical instruments. Several years before the Schumann- and Liszt-era commentators cited by Goehr, G. W. F. Hegel clearly articulated the conception in his *Lectures on Aesthetics* (1835). First, Hegel contrasted the human voice, "the sounding of the soul itself," with external instruments where "a vibration is set up in a body indifferent to the soul and its expression." But Hegel went on to explain that in highly virtuosic instrumental performance, "the externality of the instrument disappears altogether.... In this virtuosity the foreign instrument appears as a perfectly developed organ of the artistic soul and its very own property."[28] Hegel reported having experienced this effect himself in a guitar performance, where (what he described as) the tastelessness of the battle-imitative music, the ignorance of the performer, and the triviality of the instrument all faded away, as one witnessed the guitarist "put into his instrument his whole soul."

In responding to scenes of pianos going to the dump, however, commenters did not describe pianos in such prosthetic terms. Instead, they spoke of the piano as friend, family member, pet, and loved one and as having its own soul. Goehr is dismissive of such attributions of soul and vitality to musical instruments, writing that "when performers speak of pulling the energy or soul out of their instrument, what I think they really mean is that they are putting their energy into it. The way not to see a violin as an external, mechanical instrument is to see it as an extension of yourself, the violinist."[29] But the soul-possessing conception of musical instruments is equally deserving of consideration; it too is an enacted mode of understanding human-instrument relationships and no less viable a means to see a musical instrument as other than a mechanical tool. Rather than assume that only humans have souls that may be extended to or through instruments and that anyone who suggests otherwise is being silly (as our *New York Times* commenter above felt) or deluded, it is worth taking seriously the souls of instruments. Doing so reveals that despite a modern paradigm that would deny agency and animacy to artifacts, musical instruments have been a site for alternate conceptions and experiences of the relationship among souls, human bodies, and material objects. How and why keyboard instruments might be said to have souls independent of their human players is thus a significant historical and cultural question.

The intuitions of "thing theory"—one of a variety of early twenty-first-century efforts to bring critical scrutiny to the distinct beings and powers of the material world—may help explain the peculiar piano vitality that arises on the scene of the garbage dump. Thing theorists such as Bill Brown, W. J. T. Mitchell, and Bennett draw a distinction between *things* and *objects*—a distinction that separates not two different classes of stuff in the world but two different ways of encountering that stuff. Objects exist in relation to human subjects; they appear with names, identities, and functions that allow one to look through their physicality to their significance for human endeavors. Things, by contrast, assert an existence somehow apart from and beyond human purposes and knowledge. For Brown, a thing has a kind of excess, a "force as a sensuous presence or as a metaphysical presence" that exceeds the mere materiality or utility of an object.[30] Further, Brown finds that "we begin to confront the thingness of objects when they stop working for us... when their flow within the circuits of production and distribution, consumption and exhibition, has been arrested, however momentarily."[31]

Being discarded—becoming trash—is one way in which the normal flow of objects is brought to a halt. Arguing for the disruptive power of trash, Maurizia Boscagli writes that a discarded object is "dropped from the networks that give it economic and affective significance" and instead "points beyond official taxonomies of value."[32] If official taxonomies of a piano's value are monetary and artistic, the trashed piano points to alternative value-conferring networks based on kinship and possession of a soul. In losing its usefulness as a musical object, the piano asserts an autonomous metaphysical presence—a transition from nonliving to living.

Yet pianos do not fit the thing/object distinction as neatly as most objects contemplated by thing theorists, since they can also assert a metaphysical presence while they are "working" within their usual networks of value. That is, pianos may come alive in the garbage dump but also on the concert stage and other sites of their human use. So, it is perhaps not quite right to regard as fully independent the piano's two species of soul—one received temporarily from the performer, one native to the instrument. In her study of early nineteenth-century German conceptions of musical instruments, Amanda Lalonde suggests that the impression of an instrument being the extension of a player lingers beyond the time of performance, such that "in the Romantic experience, one cannot come across an instrument without calling to mind its latent resonance and the connotations of animation borne by sound."[33] Rather than needing to choose between "extension of performer's soul" and "soul-possessing" conceptions of musical instruments, they may easily blur together. In his *Musikalische Rhapsodien* (1786), for instance, the German musician and writer Christian Friedrich Daniel Schubart described the clavichord as both an extension of the player ("the soundboard of your heart," "soft and responsive to every breath of the soul") and a breathing, soul-possessing entity in its own right ("your clavichord breathes as gently as your heart").[34]

The identification of sound with animacy and inner life, so strongly present in the writings of early Romantics such as Gottfried Herder and E. T. A. Hoffmann, is also evident in the earlier tradition of harpsichord mottoes. As we saw in chapter 3's discussion of Diderot's human-harpsichord analogy, these sayings typically appeared in Latin above the keywell of instruments and were often phrased as if spoken by the instrument itself. For instance, the Latin motto *Dum vixi tacui mortua dulce cano*, which appeared on a number of harpsichords, translates as, "While living I was silent; now dead I sing sweetly."[35] Like other first-person mottoes, this one invokes a paradox; as E. K. Borthwick puts it, it describes "in riddling terms" the fact that a creature or material, silent in life, gains voice in death.[36] The paradoxical appearance of the phrase thus arises from the normal association of death with silence and stillness, sound with movement and life. The paradox deepens if we add that it is in death that the material stuff of instruments gains not only voice but also its power over human souls. It is this peculiar relation between material and immaterial that Cypess finds at the center of early modern instrumental music, summed up in a line from William Shakespeare's *Much Ado About Nothing*: "Is it not strange that sheeps' guts should hale souls out of men's bodies?"[37]

While such mottoes largely went out of fashion with the harpsichord in the mid-eighteenth century, the ability to imagine an inner being—an *I*—for domestic keyboard instruments and to relate to them as to a sort of person more than to a mere object carried on. The story of C. P. E. Bach parting with his Silberman clavichord provides a well-known example. According to Dietrich Ewald von Grotthuß, the fortunate new owner of the instrument in 1781, Bach "felt like a father who had given away his beloved daughter: he was pleased, as he himself put it, 'to see it in good hands,' yet as he sent it off, he was overcome by a wistfulness as if a father was parting from his daughter."[38] While the account describes a filial attachment between Bach and his instrument, it is also clear that gender operates here to narrow the gap between person and property; as with a daughter, but unlike with a son, Bach's parting with his clavichord takes the form of transfer into the hands of another man.[39]

Pianos take on similarly person-like status—where persons have varying degrees of agency or autonomy in relation to others—in early nineteenth-century treatises on the piano. In 1801, the piano maker Nannette Streicher published a tract for owners of Streicher instruments that counseled kindliness toward and a sense of equality with one's piano. "Just as little as he tyrannizes [*tyrannisirt*] his fortepiano," Streicher wrote of the true musician, "so little, also, is he a slave [*ein Sclave*] to it." In fact, "he knows very well how to let his instrument speak [*sein Instrument sprechen zu lassen*]," the performer-instrument relationship taking a suggestively dialogic form of a partnership or collaboration.[40] By contrast, Streicher described a pianist unworthy of imitation as abusive toward the instrument: "He flies into a fiery passion and treats his instrument as if he were one seeking vengeance, has his arch-enemy in his hands, and with horrible delight will torture him slowly to death."[41] Tyranny and enslavement—total mastery and subjugation—thus map the field of possible relations between player and instrument as extremes to be avoided. In between, both instrument and player are listeners and speakers, and the image of torturing an instrument to death implies by negative example the imperative to respect and care for instrument life.

Personification of the piano as beloved or victimized object is familiar from the reception of Franz Liszt, whose audiences found both treatments of the instrument equally exciting.[42] Such perceptions of the piano in public had their counterpart in domestic spaces; in the context of a household, the piano figured as a companion who excelled at emotional support. In 1855, a *Chambers's Journal* article on the history of domestic keyboard instruments received the title "The Story of a Familiar Friend" and concluded by describing the piano as an instrument "which deserves our truest gratitude and affection, which celebrates our happiest, and soothes our saddest hours, and to which none amongst us can refuse the name of Our Familiar Friend."[43]

Such discourses implied without plumbing the inner lives of pianos. That was left to fictional works like *A439: Being the Autobiography of a Piano* (1900). This imaginative novel was written collaboratively by twenty-five musicians under the editorial guidance of Algernon Rose, a composer and partner in the piano firm of Broadwood and Sons. The book is narrated from the perspective of a grand piano, which recounts its life starting with its initial construction, proceeding through triumphs and tribulations (including a complete rebuild after fire damage), and ending with the joy of being played by the queen herself. The first chapter, penned by Rose, describes how the piano became an individual subject, an *I*. We learn that the action—the internal mechanism that converts the motions of fingers on the keys to the motions of hammers on the strings—is the piano's brain. When the action is inserted, it allows the piano to consciously perceive the goings-on of its various parts. These parts produce a great cacophony as they complain to each other—soundboard to bridge, bridge to string, string to wrest-pin, stud, and hitch-pin—each blaming another for its own discomfort and the discomfort it causes for others. Each part thus has sentience but also interdependence within a great network ("*I* am not responsible for myself," says a string).[44] Finally, a regulator at work on the action puts an end to this internal pandemonium by installing the damper heads upon the strings. With this silencing of the piano's individual parts, the piano becomes a coherent, integral whole: "I had 'found' myself!" the piano explains. "My soul, my palpitating, sexless, breathing soul, had been evolved! Within me, even as a pearl is embedded in the guileless oyster, my spiritual self had taken up its residence."[45]

Rose continues to reflect on the piano's soul, relating it to the credit pianos should receive for their artistic work. The piano argues that "the music produced from me—by even a Liszt or a Rubinstein—could not be worth listening to, were it not for my soul, which imparts to the music its nobility."[46] People who consider musical instruments mere tools for interpreting the works of composers thus make a grave mistake. "Think of our feelings—yes, we *have* feelings—when, after enabling a player, through our glorious tone, to get through a Beethoven sonata in public without breaking down, we find him applauded to the echo, and ourselves slighted and shut down with a bang. This treatment is iniquitous: for it is the instrument which has won the success, and not the pianist."[47] The piano thus adopts the same all-or-nothing stance attributed to the pianist, merely reversing who deserves the credit rather than offering an alternative in which the performance is collaborative and the credit mutual.

The piano admits that the precise location of the piano's soul—that true source of the music's "nobility"—is difficult to define, though some say it is in a certain layer of pure silk within the hammerheads. At the time Rose was writing, the location of the piano's soul was in fact a matter of debate. For most, the piano's soul resided in the part of the instrument responsible for its tone quality, and this was thought to be the soundboard. In his *History of Pianos* (*Geschichte des Claviers*, 1868), the first part of which was devoted to the acoustics of the piano, the Leipzig Conservatory professor Oscar Paul argued for the determinative role of the soundboard in the piano's tone; the piano's strings had insufficient mass to be the source of the instrument's tone, he reasoned, and so were instead responsible for stimulating the soundboard from which the true tone was emitted.

Siegfried Hansing, technical director at the piano firm Behr Brothers and Company, saw things differently. In *The Pianoforte and Its Acoustic Properties* (1888), he noted that the soundboard had been the subject of endless studies and experiments yet remained poorly understood. That the soundboard bears chief responsibility for the piano's tone he considered a myth. Instead, he argued that the strings (including the manner in which they were struck) were primarily determinative of the tone quality, and the soundboard merely for augmenting or amplifying the sound.[48]

To identify *one* part of the piano as responsible for its tone is a quixotic goal; but my interest here is not in identifying a correct theory of piano acoustics, but rather in the fact that the ideas of a piano soul and piano tone came together to make such identification seem necessary. The desire to pinpoint a location for the soul follows in the tradition of, for instance, Descartes rather than Diderot. Descartes (in)famously identified the pineal gland as the seat of the soul. As Keiser points out, his rationale for locating the soul here was that he could not "find any part of the brain, except this, which is not double," making this the only contender for a part of the body where all various sensations (from two eyes, two ears, etc.) would be united into one "before being considered by the soul."⁴⁹ By contrast, as we saw in the previous chapter, Diderot used the harpsichord to theorize how conscious thought could arise from the materiality of the whole system; and he imagined the soul migrating throughout the human body, making whatever part in which it resided the most lively and influential over the rest.

Inability to satisfactorily explain how the soundboard contributed to the piano's tone, far from shutting down speculation about its importance, became a mark in its favor. For instance, a 1909 article on the importance of musical sound to human life included a section headed "Scientists Cannot Analyze the 'Soul.'" It explained that the rules for a good soundboard are incapable of being formulated, try though many had; only the wisdom of experience could "be depended on to fashion aright the soundboard—the soul of the piano" so as to produce something like the "famous 'Chickering tone.'"⁵⁰

While the idea that the "soundboard is the soul of the instrument" has remained the conventional wisdom, other parts of the piano have contended for status as the site of the soul.⁵¹ Knabe placed the soul in the action, an idea motivated by the significance the company wished to make of the fact that they made all their own actions, unlike other piano manufacturers that outsourced action construction.⁵² The famous nineteenth-century pianist Anton Rubinstein is said to have remarked, "The more I play the more thoroughly I am convinced that the pedal is the soul of the piano; there are cases where the pedal is everything."⁵³ Notably absent from the field of contenders for site of the soul were the keys, known metonymically by the twentieth century as "the ivories," thanks to the standardization of ivory veneers, in turn made possible by the dramatic nineteenth-century expansion of European colonial presence in Africa. The makers of ivory key veneers devoted considerable effort to transforming the grain and color of elephant tusks into smooth white surfaces, and the frequency of comparisons between skin color and piano keys lends support to Sean Murray's suggestion that "for Europeans and Asians, its resemblance to skin is perhaps its greatest subliminal attribute."⁵⁴

That white skin and soul within might have been necessary counterparts in the material and social construction of pianos places the instrument within the history not just of race and technology but of "race *as* technology," following the move by media theorists Wendy Chun, Beth Coleman, and others to think about "race not as a trait but as a tool—for good or for ill."⁵⁵ Given that keyboard instruments had selves before they had standard-issue ivory or even white-colored keys, the nineteenth-century insistence on ivory key veneers extended the ill use of race as technology for sorting persons and things, humans and machines. It also returns us to the problem with anthropomorphism—the certain humanness it requires to project it elsewhere. The "case of the piano pantaloons," as R. Allen Lott terms the question of why stories of Americans covering piano legs circulated in the latter half of the nineteenth century, illustrates the kind of inferences that follow from such anthropomorphism. Upon finding piano legs covered up with clothing fabric, British commentators saw an equation of piano legs with women's legs and interpreted the practice as an expression of excessive American modesty. American writers, meanwhile (and sometimes in defense), described covering piano legs as a functional practice intended to protect ornately carved legs from "injury."[56] Whether or not there was any truth to the British read on the practice, the anthropomorphic analogy was certainly unhelpful for thinking about how to care for a piano with materials at hand. By contrast, to personify a piano (following Barbara Johnson) is to consider its subjectivity, its rights, its coherence as a singular or aggregate entity—to consider how it might be a lyrical or legal *I*—in ways that need not resemble human form and on terms that need not bring in tow all the same considerations due to non-piano persons.

PIANO TRANSUBSTANTIATION

Wakin's story on piano dumping included a striking set of data: while only 41,000 pianos were sold in America in 2011, down from a peak of 365,000 in the early twentieth century, 2011 also saw the sale of 120,000 digital pianos and 1.1 million keyboards. Such digital substitutions and simulations put to the test what the essence of a piano is. Should one celebrate the success of these forms of the piano? Or decry their displacement of the "real" thing?

Pianos are one of many acoustic and analog instruments whose digital substitutes have been criticized as cold, lifeless, and soulless and for reducing infinite nuance and variety to fixed and finite numbers. Accordingly, most of those moved to comment on Wakin's article took little comfort in the rise of digital pianos and keyboards. Instead, the story and its reception mark a historical juncture at which discarding *a* piano (or several pianos) morphs easily into discarding *the* piano, the acoustic instrument all told. Reports of widespread piano dumping raise the possibility that the piano will go (is already going, has already gone) the way of the harpsichord and the clavichord—removed from the everyday to become a thing of the past, the province of specialists in historical performance.

The makers of digital and software pianos, on the other hand, have been working hard to establish that their products in fact preserve the soul of acoustic pianos. Synthogy, for instance, is an audio software company devoted to making "virtual" pianos, or software instruments, based on samples of acoustic pianos.[57] These samples, recorded from instruments like the Bösendorfer 290 Imperial Grand and Steinway Model D Concert Grand, are figured as "the heart and soul" of the software instrument, augmented by digital processing that performs functions like smoothing the dynamic gradient.[58] Synthogy founder Joe Ierardi equates piano sound and soul when he remarks, "I think the big knock on a lot of digital instruments has been that they have no soul... that there's this barrier that they can't create. And I think we are our most successful if people feel... like, 'Hey, what are they doing here? They stole the soul of this instrument?' I mean, I would take that as a compliment."[59] Meanwhile, the name of Synthogy's flagship product—Ivory II Grand Pianos—invokes a previous transformation in piano materiality that piano soul and sound survived: the replacement of ivory key veneers with plastic designed to replicate it. Though as with the idea of piano immortality, the stronger association in the minds of many consumers may be the more materially distanced one, to ivory as a symbol for the piano.

For others, the essence of the piano is not in the sound alone but also in the way it feels to play the instrument. Digital pianos like Yamaha's AvantGrand strive to replicate both the sound and feel of an acoustic piano. The AvantGrand resembles a baby grand but plays samples of a Yamaha CFX and a Bösendorfer Imperial. These samples are triggered using the same action as an acoustic piano, but instead of hitting strings, the hammers hit a padded bar, and optical sensors measure the speed at which the hammers pass to trigger the appropriate samples. Yamaha markets the instrument as "innovation with soul," a theme echoed by its endorsing artists. As pianist and Yamaha artist Francesco Tristano says, "It definitely has a soul. I feel the sound is coming from within. I don't know how."[60] A reviewer in *The Economist* concurred that Yamaha had captured the feel of playing an acoustic instrument. Whereas he regarded other digital pianos as "soulless digital imitations" and "lifeless imitations of the real thing," he found the AvantGrand a viable substitute for an acoustic instrument, since it recreated "all the bangs and crashes that go on inside a real piano."[61]

Software and digital pianos thus stake out slightly different positions on the essence of the piano, placing different degrees of emphasis on the physicality of the instrument. But both rely on acoustic piano sound converted into digital information, making the claims made for both exemplary of what Hayles has identified as a defining characteristic of early twenty-first-century Western culture—"the belief that information can circulate unchanged among different material substrates," such that it can "flow between different substrates without loss of meaning or form."[62] Hayles illustrates this position with the so-called Turing test, devised by Alan Turing to operationalize the question of whether machines can think. Turing proposed that if one cannot distinguish a human from a machine based on the way each answers one's questions through a text interface, then it is effectively demonstrated that machines can think. Hayles draws attention to the fact that Turing's test carefully erases embodiment from what counts as intelligence. Similarly, claims that software or digital pianos possess the souls of acoustic pianos require dispensing with not only the idea that the "soundboard is the soul of the instrument" but also with any essential relation between the materiality and the soul of a piano. In fact, in his review of the AvantGrand, *The Economist* writer imagines a Turing test for digital pianos to establish whether they equal acoustic pianos. He suggests that most would pass only under conditions of listening alone, whereas the AvantGrand would also pass under conditions of playing the instrument. Yet in both cases, any differences in piano materiality that cannot be detected under the test conditions are discounted as irrelevant.

Countering the capacity of software and digital pianos to replace acoustic pianos is a new valorization of the specifics of piano wood, strings, and metal. Where the early decades of the twentieth century saw piano materiality fade from public view, displaced by the piano as symbol of music, the early years of the twenty-first century have seen piano materiality come dramatically back into the spotlight. Films about great pianists (featuring single artists like Arthur Rubinstein, Mitsuko Uchida, or Thelonius Monk or surveying the field as in the 1999 film *The Art of Piano*) have been common since the mid-twentieth century.[63] Recently, however, a spate of movies has put the focus squarely on how pianos are made, maintained, and refurbished, making the piano documentary a veritable genre unto itself. *Pianomania* (2008) follows the work of a Steinway piano technician as he painstakingly prepares instruments for concerts and recording sessions to the artists' exacting specifications. *American Grand* (2013) and *Sitka: A Piano Documentary* (2015) focus on the challenges and rewards of rebuilding an old instrument. *Note by Note: The Making of Steinway L20137* (2007) intersperses the making of a Steinway grand in Steinway's Astoria factory with discussions with the musicians who play the instruments. One learns both about the craftsmanship that goes into a Steinway grand and musicians' appreciation for the unique "personality" of each instrument.

In a parallel set of stories in news and trade publications, musicians contend with the changing legal status of their instruments, if those instruments contain (or may contain) African elephant ivory.[64] Since 1973, the Convention on International Trade in Endangered Species of Wild Fauna and Flora (CITES) has regulated the movement of designated materials across international borders, in tandem with national laws and policies. In 2014, the US Department of the Interior announced new rules for African elephant ivory in response to "record high demand" for the material, banning all commercial import and nearly all commercial export and newly restricting its commercial exchange within the United States.[65] Complaints about the impact on internationally touring musicians led to an allowance for noncommercial import of ivory-containing instruments with proof that the ivory was legally sourced and the instrument purchased prior to the date of the new rule.[66] Such instruments are eligible for a CITES certificate, a form of documentation introduced in 2013 specifically for musical instruments and officially described as "a type of passport" that is to be presented to "the appropriate border control officer who inspects the original and validates it with an ink stamp, signature and date to show the history of movement from State to State."[67] About a century after the demand for pianos with ivory key veneers brought African elephants to the brink of extinction, the threat of that extinction has brought pianos to the brink of legally human in the frameworks of international travel and commercial trade.

In *Note by Note*, the pianist Pierre-Laurent Aimard reflects on the variety of relationships pianists have to their pianos, as well as the status of these instruments within the system of classical music performance, noting there are "people in love with *their* instrument, people who will play only with *their* instrument. Other people that fight with or against the instrument. Other [people] that consider the instrument just as an instrument. But somewhere, the goal is higher. The goal [is] the pieces and the worlds they construct. It's an ideal world that is much higher than instruments. But the instruments allow [one] to open the door, of course."[68] Piano documentaries and dumping stories, by contrast, attest to a desire to value pianos not just as instruments in service of musical works, but in their own right. Bennett has suggested that sometimes, objects have "the power to startle and provoke a gestalt shift in perception: what was trash becomes things, what was an instrument becomes a participant."[69] Pianos at the garbage dump provoke precisely such a gestalt shift, bringing into focus myriad ways of relating to pianos other than as mere tool—ways that have long been profoundly impactful but have often been reduced to incidental importance or discounted as fictional silliness. As the future of acoustic pianos remains an open question, it may not be musical works that decide the answer. It may be material factors together with humans' capacity to personify—to form meaningful relationships beyond the human.

FIVE

Genres of Being Posthuman
Chopped & Pitched

"A lot of people don't want to sound like people anymore," *Washington Post* music critic Chris Richards concluded from listening to American pop radio in the summer of 2016.[1] He was responding to a particular vocal sound. The music website Genius later identified 2016 as the year "chopped & pitched vocals took over pop music," and it was to this effect—produced by sampling and pitch shifting a singer's voice—that Richards pointed in three exemplary songs: Calvin Harris and Rihanna's "This Is What You Came For," Kiiara's "Gold," and Larsson and MNEK's "Never Forget You."[2] Using *people* and *human* interchangeably, Richards described these songs' "artificial vocal spasms occur[ring] to varying degrees of dehumanization" but added that the songs "should remind us that an artfully overprocessed vocal is always tethered to the singer's humanity" (audio 20; audio 21; audio 22).[3]

Vocal processing that sounds nonhuman is nothing new to pop radio. Since the 1970s, the vocoder and talk box have been heard on numerous hit songs and described as variously "other" voices, including robot, computer, superhuman, and alien. In the late 2000s, critics declared the "robotic" effect of hard Auto-Tuned vocals the sound of the decade. But the "chopped and pitched" sound of 2016, while related to these precedents, was also distinct from them, as was the context for making sense of them. Richards's pairing of "dehumanization" with being "tethered to the singer's humanity" points to the particular way in which music critics in the second decade of the twenty-first century navigated shifting aesthetic sensibilities in connection with shifting conceptions of being and sounding human. One can point to prior transitions in pop music history where vocal sounds marked as artificial became reinterpreted as "natural"; crooning, for instance, underwent such a transformation.[4] Rather than assimilating new sounds to an organic model of *the human*, however, the reception of chopped and pitched vocals in the 2010s entailed rethinking that model—contemplating, for instance, the degree to which humanity might include sociocognitive "upgrades" such as a cell phone or even exist purely as digital code. In other words, Richards's remarks both call up the human presumed in romantic and rock critical traditions to be an organic entity bounded by the borders of the skin and participate in a developing brand of posthumanist music criticism that envisions humans as techno-organisms—cyborgs.

The voice and its digital manipulation were prominent concerns of both musicians and critics ca. 2016. Some developments moved toward less anthropocentric, more instrumental conceptions of the voice. In the early days of sampling's mainstream adoption in recording studios (the late 1980s), Thomas Porcello found that audio engineers drew a sharp distinction between "sampling an instrument and sampling a voice—even a single sung vowel," for the reason that vocal sounds were "produced from within the individual" and were therefore unique. Instrumental sounds, by contrast, were externally produced and so more readily simulated.[5] By the 2010s, this distinction no longer seemed so pertinent. Electronic dance music (EDM)–oriented magazine *Attack* observed that the use of samplers and computers had "blurred the lines between vocal tracks and instrumentals." Working with recorded voices, dance music producers developed a variety of practices for "creat[ing] vocal-led tracks without ever going near a microphone," as well as for "us[ing] vocal samples as instruments in their own right."[6] A related shift is found in critical reception; in a 2001 article on electronic dance music and its mechanized vocals, Susana Loza wrote that although "(con)fusions of the organic with the technological are ubiquitous" in the genre, "they still sound uncanny."[7] In 2013, by contrast, Auner argued that sampled voices had gone "from the uncanny to the unremarkable," noting "how normal . . . it is becoming to hear and accept voices that have only distant relationships to the bodies that produced them."[8]

While becoming more commonplace, sampled and manipulated voices also took on newfound importance, thanks to what prominent electronic musicians characterized as a crisis of musical anonymity and homogeneity brought on by common access to the same tools of music production. As Thomas Bangalter of the duo Daft Punk remarked in a 2013 interview, "In electronic music today, there's an identity crisis. You hear a song: Whose track is it? There's no signature. Everyone making electronic music has the same tool kits and templates."[9] In a revival of the logic of Porcello's engineers, voices emerged in the mid-2010s as a solution to this problem of sonic signature. As DJ/producer Diplo explains, "Anybody can copy any synth now. You can find the synth sounds; you can find presets. But if you manipulate vocals, it's something really original."[10] Similarly, DJ Snake explains, "I don't really use synths, I use my voice for the drops and that's kinda like my style, my signature... my touch. People recognize me because they hear some weird noises in my music and most of the time that's just my voice."[11] Even as voices were interchangeable with synthesizers, there was also an investment in the unique sound of individual voices and their ability to betray the particular human(s) responsible for a track.

This sociotechnical context makes the question of how voices relate to the bodies producing and manipulating them all the more significant for how one hears and interprets songs like the ones singled out by Richards. That question is more complex than Richards's pithy assessment that "a lot of people don't want to sound like people anymore" would imply. This is not only because the listener plays an active role in linking voice and body, with results shaped by beliefs about race, gender, sexuality, and the human in ways scholars such as Suzanne Cusick, James Davies, Nina Eidsheim, Ana María Ochoa Gautier, Alisha Lola Jones, Jennifer Lynn Stoever, and Kira Thurman have brought to light.[12] It also has to do with who was involved and how in making the music in question. All three of Richards's music examples feature women's voices that are chopped and processed, while the producers of these tracks—the ones doing the chopping and processing—were men. Richards glosses over this gendered division of labor, but it is worth considering when inferring whose and what desires the tracks' vocals make audible. He also elides aesthetic differences between his examples, creating a single human-machine dynamic where one could instead hear different configurations of humans and machines, different "genres of being [post]human," to take up Sylvia Wynter's phrase to forestall making one mode of being "isomorphic with the being of being human itself."[13] Richards's comments thus provide a jumping-off point for examining the rise of posthumanist sensibilities in pop music and music criticism and what three pop hits of 2016 say about being posthuman.

SOUNDS LIKE DEHUMANIZATION

In popular music discourse, the term *dehumanized* has often been used to describe the effect of sounds associated with the mechanical or machine-made, such as a metronomic beat or robotic voice. It marks such musical features as deviations from sounding human in the sense of loss, the subtraction of human qualities deemed both essential and desirable—the foremost qualities in these musical contexts being emotion, soul, and free will. The concept thus encompasses basic assumptions about what it is to be human and what is capable of corrupting that humanity. As we saw in chapters 1 and 2, the term *dehumanize* carried a different range of implications in the early Enlightenment, turning a human into a machine not being prevalent among them; and the early reception of musical androids was marked not by anxiety around machines being able to perform like humans but by pleasure at imagining another being's sensibility—a pleasure that affirmed one's own status as a feeling human. After about 1750, however, the tide began to turn toward needing to distinguish human performance from its machine simulations and toward conceptualizing machines as what humans should not be. Assessing professional musicians' reactions to Vaucanson's flute-playing automaton in the 1750s, David Yearsley identifies a newly crystallized assumption that "a machine could never be endowed with the subjectivity necessary for meaningful musical expression," as well as another line of argument that emphasized rationality as an exclusively human faculty necessary for musical creation.[14] In her study of the manufacture and reception of late eighteenth-century keyboard-playing women automata, Voskuhl notes, "When we observe [them] playing music, we not only are unsure whether they can communicate affects to the audience (that is, us) as a human musician would. We are also not sure whether the automata are in a self-reflexive relationship to their own sentimentality."[15] Rather than an occasion for pleasurable imagining and feeling, such ambiguity became a source of anxiety and occasion for drawing a distinction between merely mechanical aspects of music making and those aspects requiring a thinking and feeling soul.

By the mid-nineteenth century, the idea that musicians might be or turn into machines was pervasive. "If I were younger, I might even launch myself as a machine," pianist-composer Frédéric Chopin wrote from Scotland in 1848, envisioning a nonstop touring schedule playing "tasteless" music for money, "but now it is already difficult to begin to make a machine out of myself."[16] "I feel like a song machine," wrote opera singer Pauline Viardot in 1859, likewise about performing night after night in Great Britain: "At 8:00 in the evening they push my spring and I sing."[17] To sing or play mechanically was also a standard criticism leveled against musicians who failed to convincingly convey an interior being of emotion and intelligence. As Céline Frigau Manning found in her study of nineteenth-century discourses on singers, mechanical figures such as automata, puppets, and marionettes were "usually used satirically to condemn specific singing or acting practices."[18] The common implications of these figures were a lack of feeling, of soul, and of intelligence. One singing practice so condemned was vocalizing without clear declamation. An Italian music critic complained of proliferating *"vocalise machines"* on the opera stage, singers whose bodily movements did not "reveal a *soul*" and whose singing was marked by an indifference to "*consonants*, which is another way of saying that they do not care about making themselves understood."[19] Because articulate speech was a sign of intelligent thought, its absence could be taken to indicate a mindless singing machine. For some critics, the mindless, soulless singer was a particularly female phenomenon. As the French composer Hector Berlioz complained in 1845, "We find a fair number of female singers, popular from their brilliant singing of brilliant trifles... they have voices, a certain knowledge of music, and flexible throats; they are lacking in soul, brain, and heart."[20] The ready conflation of female performer and automaton was consistent with a widely documented and enduring suspicion of whether women were capable of musical creativity at all or were at best technically proficient vessels for male creative genius.[21]

In the latter half of the twentieth century, the cyborg introduced an alternative to the purely mechanical automaton or robot. A hybrid being, the cyborg (a portmanteau of cybernetic organism) emerged from the post–World War II field of cybernetics. Along with the brain-computer analogy discussed in chapter 3, the field developed the view that "the human and the machine might be regarded not as wholly distinct entities subject to comparison, but rather as interconnected parts of a single integrated system."[22] Of the difference between robot and cyborg figures, Claudia Springer has observed: "While robots represent the acclaim and fear evoked by industrial age machines for their ability to function independently of humans, cyborgs incorporate rather than exclude humans, and in so doing erase the distinctions previously assumed to distinguish humanity from technology."[23] Initially conceived to describe a man able to function autonomously in space thanks to self-regulating machinery, the cyborg was reimagined by Haraway as both a description of humans in the late twentieth century—"We are all chimeras, theorized and fabricated hybrids of machine and organism"—and a figure of social transformation by way of dissolving traditional dualisms.[24] In particular, the cyborg for Haraway and subsequent cyborg theorists signaled thoroughly ambiguous or breached boundaries between human and animal, organism and machine, physical and nonphysical, and related dualisms such as natural and artificial.

Despite its differences from robots in theory, in ca. 2000 music criticism, the cyborg often functioned as a synonym for robot, with the same valences of deficiency and distance from the "real." Consider two examples from the *Village Voice*, the New York City–based alternative weekly renowned for its serious intellectual engagement with popular music.[25] In 1999, Jane Dark, a pseudonym of the poet Joshua Clover, whose persona was that of a "catty" "wise-ass" teen girl, cast Mariah Carey as a "singing machine."[26] After name-checking Haraway to observe that Carey's rise coincided with the publication of *Simians, Cyborgs and Women*, Dark proceeded to offer unflattering characterizations of Carey's technicity. Saying that she offers a "vision of love that only dogs and computers could hear," Dark implied an exclusion rather than incorporation of humanity. Similarly, by suggesting that like Rachael, the replicant in the film *Blade Runner*, "she probably thinks she's human," Dark wryly denied Carey a human element.[27] In 2007, Tom Breihan cast Rihanna as "a singing cyborg" only to write with similar sarcasm about Rihanna's *Good Girl Gone Bad*: "Given that Rihanna's voice is a blank, icy glass-shattering yowl, this album is a really good look for her; it's always just one working part in a very expensive and elaborate engine." The dismissive turns of phrase Breihan uses to describe Rihanna provide the prelude for his complaint that Rihanna receives massive hype, whereas an artist like Amerie—who "makes sure we can hear her voice strain" over samples arranged "in ways that feel a lot more organic than most R&B"—is largely ignored.[28]

Targeting different vocal stylings—Carey's virtuosity and extraordinary range, Rihanna's "thin" voice—both Dark and Breihan deploy high-tech and animal imagery to convey an absence of human qualities. While Dark does so tongue in cheek, the result is nonetheless a portrait of the artist as a machine; Breihan is more clearly invested in the aesthetic values of organicism and a voice in which one hears the fleshiness of a body. That both Carey and Rihanna are women situated in the Black music genre of R&B suggests that their mechanical reception has to do not only with the historical tradition of likening female performers to machines but also with expectations that their singing conform to a particular kind of "soulful" voice—perhaps sounding more like gospel-trained Whitney Houston, whose musical craft was often mythicized by characterizations of her as a goddess or angel, vocal spirit rather than technology.[29]

Auto-Tuned voices were ripe for the application of a human-versus-automaton critical apparatus. "You can only feel so bad for a robot," Sasha Frere-Jones wrote in 2008. He was explaining why on Cher's 1998 hit "Believe," which pioneered the Auto-Tuned vocal sound, the effect was reserved for phrases on the verses, whereas the chorus ("Do you believe in life after love?") unleashed the power of Cher's "human" voice.[30] The characterization *robotic* supported critics' contentions that Auto-Tune was incompatible with emotional expression and, in cases where an artist did not have a track record that clearly proved their singing abilities, a cover for lack of vocal talent. Rapper T-Pain and pop stars Britney Spears and Kesha were among the artists who suffered from the equation of Auto-Tuned vocals with musical deficiency. As Ben Westhoff summed up the critical reception of T-Pain, whose heavily Auto-Tuned vocals helped popularize the effect, "You probably know him as the guy with the robot voice... accused of homogenizing music, dumbing it down, and destroying its humanity."[31] Spears and Kesha have both been the subject of viral videos purporting to reveal them singing horribly out of tune without the assistance of Auto-Tune software.[32]

Critics characterized the popularity of Auto-Tune as an epidemic or robot takeover—something running beyond human control. A 2008 video featuring T-Pain and his "vocoder" (a common misidentification of Auto-Tune) humorously dramatized the idea of the robot wresting control away from humans. The video parodied *2001: A Space Odyssey*, the Kubrick film in which HAL, the AI running a spaceship, decides to eliminate its human crew for its own self-preservation. After T-Pain asks for "less vocoder" because he's "soundin' way too much like a robot," the vocoder commandeers the recording session, ultimately making T-Pain record the line, "Thank you, vocoder, you are wonderful and have helped me tremendously" as alternate lyrics to his hit song "Chopped N Skrewed" (audio 23).[33] Yet for many critics and musicians—particularly those steeped in rock aesthetics—Auto-Tune was no laughing matter. Its machine perfection was the enemy of human musicality, as rock musician Dave Grohl expressed at the 2012 Grammys: "The human element of making music is what's most important.... It's not about being perfect.... It's not about what goes on in a computer."[34] In 2013, Lessley Anderson captured the persistent sense that an essential humanity was being lost to machines: "As humans, we crave connection, not perfection. But we're not the ones pulling the levers. What happens when an entire industry decides it's safer to bet on the robot?"[35]

"IT'S HOW PEOPLE REALLY SING"

Anti-Auto-Tune sentiment presumed some notion of how voices should sound, which critics and musicians sometimes articulated by appealing to how *people* sound. Clarifying the difference between sounding like a human or a robot, Frere-Jones pointed to the slides between pitches that Auto-Tune eliminated: "Portamento is a natural aspect of speaking and singing, central to making people sound like people."[36] At the 2009 Grammy awards, indie rock band Death Cab for Cutie carried out an anti-Auto-Tune publicity stunt by wearing blue flag pins. As lead singer and songwriter Ben Gibbard explained, they wanted to "raise awareness while we're here and try to bring back the blue note"—a pitch inflection named for its prominence in blues music. It's "the note that's not so perfectly in pitch and just gives the recording some soul and some kind of real character. It's how people really sing."[37]

Riskin has observed that "understandings of machines and of humans have... shaped one another in [an] ongoing dialectic" since the advent of human-simulating automata in the eighteenth century—a pattern well illustrated by the reception of Auto-Tune.[38] How these critics defined the *natural* or *real* was shaped by the sound of Auto-Tune; both Frere-Jones and Gibbard selected something excluded by Auto-Tune—portamento or blue note—to characterize how people sound. The mutual construction of sounding like a person and like a robot in the specific context of Auto-Tune is thus transfigured into universal standards for sounding human.

Gibbard's chain of valorized terms—the *blue note, soul, real character*—additionally sets natural singing against the threat of Auto-Tune by invoking the racialized aesthetics of rock 'n' roll. As Kodwo Eshun, Robin James, Jack Hamilton, and others have discussed, rock discourses positioned Black male blues artists as in touch with nature and masculine embodiment, uncontaminated by commercialism, and thus representative of the "authentic" and "real" that White rock musicians sought to emulate.[39] Technology, and the wider forces of modernization, posed a threat to this imagined natural musicality. This configuration of human musicality and technology extends, in turn, colonial fantasies of the "noble savage." As political scientist Sankar Muthu aptly characterizes, "noble savage" discourses performed "the dehumanization of natural humanity"; in celebrating so-called primitives as the "most purely human," these discourses also figured them as "instinct-driven and mechanical animals" whose praiseworthy features were beyond their control. By contrast, the modern writers and readers about these figures understood themselves to be humans with cultural agency that enabled them "to consciously and freely transform themselves and their surroundings."[40] Auto-Tune reception thus not only exemplifies the dialectical nature of understandings of humans and machines; it also exemplifies how claims about how people sound and about the effects of using technology have been shaped by racialized conceptions of what is natural or artificial to humans.

"PEOPLE DON'T WANT TO SOUND LIKE PEOPLE ANYMORE"

Against Auto-Tune panic, some music critics began to deploy the resources of cyborg and posthumanist theories. Seeking to appreciate rather than deprecate high-tech artifice, these critics developed nonantagonistic understandings of human-machine relations. To do so, they drew from a range of sources, which included theoretical literature, pop culture (fiction, film), ideas filtered through the musicians they wrote about, and the sounds of their music.

The beginnings of posthumanist music criticism, however, preceded Auto-Tune and centered on African American musicians animated by a need to reframe their relationships to technology and science fiction in counterpoint to racist denials of their technological interests and abilities, as well as to their general absence from sci-fi futures. The concern was thus with what Weheliye has identified as "a specifically Black posthumanity."[41] Exchanges in the early 1990s between Greg Tate and Mark Dery at the *Village Voice* and Mark Sinker and David Toop at *The Wire*—an antimainstream, proinnovative music magazine based in London—initiated conversations in this direction. Eshun, who was at the time a regular contributor to *The Wire*, offered the first extended publication on music and the posthuman with his 1998 book *More Brilliant Than the Sun*. Eshun rejected "humanist" (that is, rock critic) approaches to Black music, which he characterized as overemphasizing lyrics and sociohistorical context, celebrating soul and the authenticity of "the street," and systematically overlooking Black engagement with electronics and computers. Moreover, he argued for understanding *human* as a "treacherous category" that excluded enslaved people and their descendants in the United States—as evidenced by the fact that African Americans still had to fight for their civil rights in the 1960s.[42] In a marriage of Afrofuturism to McLuhan's theories of how media affect the sensorium, Eshun developed an account of why "postsoul" musicians (in electronic jazz, hip-hop, and electronic dance music) abandoned the human and of how the machine organization and sounds of their music reconfigure perception, subjectivity, and embodiment.[43] Against cyberculture critics who saw the disembodiment of humans by technology, Eshun argued that from a sonic perspective, "the posthuman era is... a hyperembodiment" in which "sound machines make you feel *more* intensely, along a broader band of emotional spectra than ever before."[44] Though Eshun noted that vocal processing (primarily by vocoder) creates "nonhuman subjects," voices play only a minor role in Eshun's account. The focus is on rhythmic and sonic attributes of turntables, samplers, synthesizers, and drum machines—on how humanly impossible grooves and synthetic sounds reprogram consciousness, replacing the singular *I* with a self that is multiple and divergent, and of which machines are a part.

Like Eshun's book, posthumanist music criticism in the ca. 2010s popular press often took an oppositional, revisionist stance. But, not aiming to correct the record on Black music, it was not so deeply skeptical of the human as a "treacherous category" for its history of race-based exclusions. Rather, animated by the critical backlash against—and human-versus-robot framing of—Auto-Tune, posthumanist music critics deployed the cyborg and related cultural theories to redeem technology's presence as both aesthetically valid and compatible with being human. In a 2009 article in London-based *Frieze Magazine*, DJ and culture writer Jace Clayton countered the "vocal purists [who] hate Auto-Tune" by citing Haraway's cyborg manifesto: "Auto-Tune's creative deployment is fully compatible with [Haraway's] 'argument for pleasure in the confusion of boundaries and for responsibility in their construction.'" Against critic Jody Rosen, who suggested in *Slate* magazine that T-Pain used Auto-Tune to "impersonate a computer," Clayton argued that his vocals were "quite literally... the sound of voice and machine intermodulating," a "human-machine duet."[45] In her book *Good Booty* (2017), music critic Ann Powers elaborated a cultural appreciation of Auto-Tune she had begun to develop as early as her 2008 review of Kanye West's *808s and Heartbreak*.[46] Drawing support from cyborg theorists including Haraway, Sadie Plant, and Andy Clark, Powers transformed the liability that was Spears's thin voice into a virtue, praising "its metallic tone perfectly suited for manipulation—Spears presented from the beginning as a hybrid."[47] Whereas Dark invoked *Blade Runner* to suggest Mariah Carey "probably thinks she's human"—characterizing her as deluded and disempowered—Powers invoked the optimistic motto of the film's Tyrell Corporation to describe Spears positively as "more human than human."[48] Of T-Pain, in contrast to earlier discourses that either blamed him for ruining music or credited his popular success to his technology rather than to him, she wrote that his "accomplishment was the creation of giddy confusion between the flesh and mechanics, calculation and emotion."[49] Declaring Auto-Tune "indisputably the sound of the 21st century so far" in 2018, Simon Reynolds rebutted every criticism that had been leveled at Auto-Tune, including by making the case for all singing as "introjected technology" and highlighting the rise of recording with Auto-Tune on (as opposed to applying it afterward), such that "the true voice... is Auto-Tuned right from the start." Thus, artists like the rapper Future "are almost literally cyborgs," and "we have grown used to connecting *machines* and *soulfuness*."[50]

Several common elements characterize such early twenty-first-century works of posthumanist music criticism that defend technologically manipulated voices: an assumption that technology is part of being human; a related absence of anxiety about technology displacing humans or taking over; a rejection of natural/artificial as a meaningful distinction, artifice being considered in the nature of human relationships, self-presentation, and musical performance; and finding confusion around human-machine boundaries pleasurable and therefore praiseworthy. Overall, there is a sense that technology is not only ubiquitous and necessary for human life, but also that it tends to enhance human expressivity and social connectivity. In this discourse, the cyborg represents something neither radically Other nor radically revolutionary; it is a clearer, truer representation of what *we* humans (regardless of race and gender) have been all along.

Richards's article on the sound of pop radio in the summer of 2016 sits within this broader uptake of cyborg theory in music criticism, as well as in the midst of a decisive crossover of EDM into Top 40 (as a *Pitchfork* music critic observed in February 2017, "The chasm that once separated mainstream electronic music and Top 40 has all but vanished").[51] Richards's mid-2010s writings reveal making sense of intimate human-technology relationships to be a recurring theme, with the summer of 2016 marking a transitional state in his incorporation of posthumanist theories into his listening. In 2015, for instance, Richards interviewed electronic musician Herndon, whose album *Platform* makes extensive use of glitchy, stuttering, pitch-shifted, and otherwise processed vocals. In the interview, Richards expressed skepticism toward both technophobia and techno-utopianism with the remark, "I like that you're overturning the idea that technology is corrupting our humanity—but not completely." Herndon's reply—"I really think [technology]'s just part of being a human being"—discredits the "technology is corrupting our humanity" idea by refusing a pre- or non-technological image of humans, instead positing a cyborg ontology.[52]

In a 2017 article, Richards brought a variety of posthumanism to bear on music whose artists were not explicitly engaged with such theories, arguing that the summer's EDM-pop hits thematized transhumanism. To support this contention, Richards cited the recently published *To Be a Machine: Adventures Among Cyborgs, Utopians, Hackers, and the Futurists Solving the Modest Problem of Death* by freelance journalist Mark O'Connell. In the book, O'Connell examines the mind-uploading, meat-body discarding dreams of futurists such as Hans Moravec and Ray Kurzweil and the work of scientists trying to make such dreams reality. O'Connell thus identifies the disembodied mind as the highest ideal of the transhumanist movement, and Richards used the book to suggest that the highly processed vocals on three EDM-pop hits were exemplary of this desire for "'a total emancipation from biology itself.'"[53] In an echo of Haraway's call for "pleasure in the confusion of boundaries"—and in the tradition of using language as the site for discerning the presence of intelligence—he also wondered whether "Bad Liar" had Selena Gomez singing lyrics by a human or a buggy algorithm, concluding "it's hard to know for sure, and the pleasure is in the not-knowing."[54]

Richards's summer 2016 article finds him venturing in this posthumanist direction. On the one hand, his claim that "a lot of people don't want to sound like people anymore" relies on the assumption that sounding *like a person* means sounding like a human whose body ends at the borders of the skin—a body that has not been meddled with technologically. On the other hand, he frames *not sounding like a person* as a motivated choice rather than a mark of deficiency. Further, where *dehumanization* invokes a critical tradition of framing technological meddling in the negative terms of loss, Richards's coupling of *dehumanization* to *humanity* offers a redemptive turn. His claim that vocals *always* remain tethered to the singer's humanity invalidates robot readings of vocal processing, at least when it is "artfully" done; even while dallying with dehumanization, singers remain indelibly human. The threat of technology to humanity is thus neutralized by a fundamental confidence in the persistence of humanity, symbolized by the still-recognizable human voice.

GENDERED DIVISION OF LABOR

The processed vocals of 2016 pop thrived in systems of music production, media representation, and consumption that complicate Richards's attribution of "people not wanting to sound like people anymore." While dozens of male popular musicians used the vocoder or talk box from the 1970s to mid-1990s, the popular female musicians who did so can be counted on one hand, the most prominent being Carlos, Laurie Anderson, and Hendryx.[55] Women who played with robot or cyborg personas in these decades, such as Grace Jones, largely did so without vocoder-like effects.[56] In the wake of Cher's 1998 hit "Believe," by contrast, highly processed vocals became something many female artists tried out or made a feature of their sound (examples include Madonna, Kylie Minogue, 702, Spears, Kesha).

In a 2001 article, Kay Dickinson observed the sudden shift in the gender distribution of processed voices in the wake of Cher's "Believe."[57] However, due to the misdirection of Cher's producer Mark Taylor, her analysis was uninformed by the technological impetus for this shift. Taylor told *Sound on Sound* that Cher's vocal was vocoded using a DigiTech Talker pedal, making the effect technologically continuous with earlier vocoder use. Only later did it become common knowledge that Taylor had used Auto-Tune software released by Antares the year before.

The technological differences between the vocoder and Auto-Tune clarify the dynamics of the sea change after 1998. The vocoder—like the talk box—required playing the pitches on a synthesizer or electric guitar while talking, meaning that the "vocalist" and technology user were typically one and the same person. In live performance, this integration of roles was on display, the "singer"—Herbie Hancock or Troutman, for instance—being seen playing a synthesizer while they shaped words with their mouth. Auto-Tune, by contrast, was digital processing designed to be applied to the voice after it was recorded. It enabled splitting the roles of singer and software user; now it was easy to have a singer record vocals and a producer handle the vocal processing. This division of labor—the removal of the singer from the technology—is clear in Taylor's account of manipulating Cher's vocals on "Believe": "Basically, it was the destruction of her voice, so I was really nervous about playing it to her! In the end, I just thought it sounded so good, I had to at least let her hear it—so I hit 'Play.' She was fantastic—she just said 'it sounds great!,' so the effect stayed" (audio 24).[58]

Dickinson argued that though "it would not seem untoward to derive extremely disempowering readings from male producers chopping chunks out of women's performance," there was a female-empowering reading to be made from the fact that Auto-Tuned vocals became known as the "Cher effect," with technological adeptness being credited to the female performer.[59] Yet, twenty years later, this image of female technological adeptness had not changed the reality of technology being primarily a male preserve within the pop music industry. An analysis of the 600 songs on the Billboard Hot 100 year-end charts from 2012 to 2017 found only 2 percent of producers were women.[60] Behind such statistics are stories of women being routinely doubted or thwarted in their production abilities. In a 2016 interview, the electronic musician Grimes (Claire Boucher) described working at a writing camp—a label-funded session at which producers and songwriters are brought together to come up with hit-making material: "I wasn't allowed to touch a computer... even though guys in the studio were allowed to. I obviously know how to use a computer and I know how to produce, but I had to tell the engineer what to do if I wanted to do anything." Clarifying that she is "way more of a producer than a vocalist," so it seemed unlikely she was invited only to do vocals and top line, Grimes concluded, "It was pretty gendered... all the guys made beats and all the girls did top line... there were no girls doing beats and there were no guys doing top line."[61] Grimes's experience attests to how rigidly gendered the division of labor between production and singing can be in core professional spaces of the popular music industry.

A gendered division of labor can also be seen in the conceptualization of Auto-Tune software. In her study of 1990s music criticism, Emma

Mayhew found that critics commonly articulated a dichotomy "between the emotional expression of the artist and the technical objectivity of the producer," which mapped onto female artists and male producers.[62] This same dichotomy is evident in Antares's language about Auto-Tune. As the user manual for the software explains: "Many of our most celebrated entertainers spend hours in the studio doing retake after retake, trying to sing expressively and in tune. Afterwards, their producers spend yet more time trying to correct intonation problems using inadequate tools. Auto-Tune is dramatically changing all of that. Because of Auto-Tune, sessions can focus on feeling and expression."[63] Setting up a conflict between singing expressively *and* in tune, the user manual suggests that the technical skills involved in "in tune" singing can be entirely outsourced to the producer, leaving the singer to be pure emotion and expression. That pitch control might also be a resource for expression (through portamento, for instance), whether at the discretion of the artist or the producer, is absent from this framing.[64]

The gendered division of labor in music production forms an important context for the songs selected by Richards to exemplify the *not sounding like people* phenomenon. These songs are the fruits of diverse collaborative situations. Pop star Rihanna (Robyn Fenty) first collaborated with Harris (Adam Wiles) in 2011 for her album *Talk that Talk*. In an unusual move, Rihanna agreed to credit Harris as a featured artist on "We Found Love" for his production work, a designation usually accorded a guest vocalist. When the song became a hit, it catapulted the comparatively little-known Harris to a new level of mainstream popularity. "This Is What You Came For" reunited the pair in 2016 but reversed the credit lines: Rihanna is a featured artist on the track, released under Harris's name. Songwriter Kiiara (Kiara Saulters) was unknown to the public and had as yet no performance career when her debut single "Gold" was released and became a hit. Snow (William van der Heyden)—who had previously worked with SZA (*S*, 2013; *Z*, 2014) and Selena Gomez (*Revival*, 2015)—produced the track, receiving the conventional behind-the-scenes producer credit. Larsson and MNEK both started their musical careers young—Larsson won a Swedish-television talent show in 2008 at age ten and signed with a label four years later; MNEK (Uzoechi Emenike) started racking up cowriting and production credits in 2010 at the age of sixteen and within four years worked with veteran stars Kylie Minogue and Madonna. The collaboration on "Never Forget You" came with Larsson and MNEK both being seasoned professionals yet still early in their rising careers; both sing on the track, which was coproduced by MNEK and Astronomyy (Arron Davey).

Despite the range of professional circumstances behind these tracks, in all three cases, the producer-singer relationship is very much like the one described by Taylor for Cher's "Believe". Harris's account of playing the finished version of "This Is What You Came For" for Rihanna reflects her absence from the production process after recording her vocals: "I'm kind of nervous to play it to her because I changed so many bits from when she heard it even, like musically... and she was into it."[65] Foregrounding his creative role, Snow explained that "Gold" sounded nothing like anything Kiiara had made before, and "that was the point—to make her forget what a creative comfort zone was."[66] For her part, when asked about the chopped vocals on "Gold," Kiiara professed ignorance and ultimately claimed to be mystified by the producer's work: "So the producer came in [and] just went to the computer and chopped up throughout the whole song.... I don't know, he probably reversed stuff, I don't know he's just like a genius with it so I don't know what he did."[67] In an interview between MNEK and Larsson, Larsson asked MNEK when he started producing, drawing a distinction between production versus singing that is arguably more reflective of her different lifelong access to the two skill domains than it is of different amounts of time spent learning and practicing: "Producing you have to actually spend time learning to produce. I just sing; I don't need to practice it for hours and hours."[68]

The division of labor between singing and production means one cannot assume that a voice directly expresses how a singer wants to sound. In a 2015 interview, Herndon drew attention to how the sound of female voices might in fact reflect male producers' desires: "The majority of female voices that we hear on the airwaves today have been produced by males—they go through that filter of what an idealized voice should sound like—so I think it's empowering to make that whatever I want it to be, whether that be beautiful or ugly or harsh."[69] A particular "fondness for female voices singing in their upper register" was noted in a 2016 *Pitchfork* review of an exemplary EDM-pop album (Flume's *Skin*), with *feathery* being a recurring adjective for the light and airy vocal sound that alternately soars lyrically above more abrasive, weighty synths and becomes another electronic instrument (as at the end of "Never Be Like You," audio 25).[70]

Nor can one assume, however, that the vocalists from Richards's three examples wanted to sound like people and were made to sound otherwise by producers regardless. Rather than suggesting they had no control over their sound, the comments of these singers and producers suggest the women chose sites other than production software on which to exert their agency. Recounting how their first collaboration came to be, Harris said Rihanna approached him when the two were on tour together in Australia, wanting "to do something that kind of sounded like what I was playing."

When he produced "We Found Love" soon thereafter, he "wasn't shooting for someone as high and amazing as Rihanna" to sing it, "but she heard it and she wanted it." Harris concluded: "She makes her own decisions, I think you can see that."[71] The dance floor hookup narrative of "This Is What You Came For," their second collaboration, reads like an allegory of their hit-making procedure: Rihanna is the star "everybody's watching," who catapults a song to mega success, but "she's looking at you," the producer she chose from a crowded field. Kiiara, discussing her authority over her work in an interview, noted, "Ultimately if I like it... I'm never gonna put something out I hate." Of her music videos, she remarked that for the song "Feels," "I just let the director take control," but she also acknowledged that she found such relinquishing of control difficult, "unless I really trust the person or their vision."[72] Though not asserting direct control over production, then, these women are selective about their collaborators and retain veto power over work appearing under their name.

Such is the context for inquiring how adequately Richards's claim that "a lot of people don't want to sound like people anymore" accounts for an aesthetic of "chopped and pitched" vocals that is applied by male producers preferentially to female voices. Clearly there is an entrenched gendered division of labor contributing to the predominance of male producers, and studies of women's underrepresentation in music technology professions can help illuminate it.[73] Attending to what the producers of Richards's example tracks say about voices and how music videos represent the organicism or technohybridity of human bodies illuminates how the gender disparity heard on the radio and seen in the studio relates to sounding posthuman.

"THEY'RE ALWAYS AT THE PERFECT FREQUENCY"

One reason EDM producers give for working with female vocals is that female voices are better than male voices. In 2012, Harris told National Public Radio (NPR) that despite working with his voice in the studio, "There's still a very limited range there, which is why I like working with other people—especially female singers because they suit dance music a lot better. There's only so much you can do with a male voice in dance music."[74] Speaking to the UK-music magazine *NME*, he elaborated on what made female vocals better suited to dance music: "They're always at the perfect frequency to play in a club. A good, soaring, high-mid female vocal bounces off the walls nicely, and it doesn't interfere with the bass or drumbeat. It's basic science. A man's voice is likely to interfere with the bassline, which is why you don't hear many classic dance tracks with a male lead."[75] Likewise, Snow told an interviewer in 2015, "If you're to be a guy and sing, you can only sound like a girl, it's the only way it works."[76]

By appealing to "basic science," Harris grounds the preference for female vocals in electronic dance music in nature, in physical and biological necessity. Basic science, however, fails to support Harris's claims, since men can sing well above the bass range. As Cusick has observed, the physical changes that grant boys access to the tenor or bass vocal registers do not concomitantly shut off access to vocal registers that boys share with girls.[77] To sing in this shared register, however, is to "sound like a girl" rather than a man, according to Snow's comment. Harris and Snow thus set up a two-sex system in which singing "naturally" belongs to women; to the extent that men participate in singing, they either abandon their masculinity or compromise the music.

Harris's efforts to distance himself from singing have been deliberate, as well as equivocal. On his first two albums, *I Created Disco* (2007) and *Ready for the Weekend* (2009), Harris was the primary singer, establishing himself as a "producer and vocalist."[78] The music press, however, had little good to say about Harris's singing. *Pitchfork* described his singing, which was often reminiscent of disco falsettists such as the Bee Gees' Gibb brothers, as "kitschy falsetto," a criticism that may be read as gender policing of vocal ranges.[79] In 2010, Harris announced he was quitting the vocalist role, telling the Melbourne-based *Herald Sun*, "I'm going to focus more on producing and DJing and zero percent of my time will go on singing"—though also, "I'll do some studio singing for sure, but never live."[80] His next album, 2012's *18 Months*, found him singing sparingly and reaching new levels of commercial success with Rihanna's vocals on "We Found Love."

In 2014, Harris reembraced singing, citing the encouragement of producer-vocalist Pharrell Williams, who told him (in an echo of the Antares Auto-Tune user's manual) what matters about the voice is not pitch but "getting the emotion and the feeling and the character across."[81] Thereafter, Harris said, "I started looking at myself, like my voice, differently"—and indeed, in this phase of his career, he used his voice differently. Harris no longer sang falsetto. Nor did he chop or repitch—for audible aesthetic effect—his own voice (as he did on earlier tracks like 2007's "Neon Rocks" and "This Is the Industry" and 2009's "Limits" and "Stars Come Out," and as he still did with his collaborators' vocals). Instead, with his own voice, he went for a relaxed delivery, gravelly timbre, and unprocessed sound. The arrival at this seemingly casual, unstudied vocal style was the result of extensive experimentation and practice. As Harris told an interviewer who expressed surprise that this was the voice of "a white kid from Scotland"—implying a mismatch between this identity and the body the

FIGURE 5.1. Rihanna's digitality in the music video for "This Is What You Came For" (2016).

voice conjured—"twelve years of trying to make my voice sound good in the studio, I finally learned how to do that."[82] For the song "Summer," Harris described coughing for a whole day in order to obtain a rough vocal quality.[83] The result is a vocal that foregrounds the fleshiness of its source and refuses to disperse into or become interchangeable with its electronic surroundings. Allowing no challenge to the body's integrity, this voice affirms its masculine origin. The result of Harris's "looking at myself, like my voice, differently" is a switch from using his voice as a malleable instrument to deploying it as an expression of his (organic, manly) self.

Gender differentiation thus takes place not only along the dimension of vocal register but also along the dimension of the body's availability for electronic manipulation. Just as female singers' voices undergo more obvious digital processing, so too do their bodies as represented in music videos. The video for "This Is What You Came For," for instance, shows Rihanna in various states of digital dissolution, as she seems to dance in and out of a cube whose sides are projection screens (fig. 5.1). The pixelated, glitchy appearance of Rihanna's body is reminiscent of a visual effect applied to Cher in the music video for "Believe," which synchronized digital fragmentations of Cher's body with the audible glitching of her voice. By contrast, the music video for "Summer" presents a firmly bounded, organic image of Harris. Harris is primarily shown in a white T-shirt and jeans, walking purposefully toward the camera down the middle of a wide paved road that cuts through a rugged landscape (fig. 5.2).

Adding to this differential availability of the body to digital processes are gendered conceptions of player-instrument relationships. Examples of this phenomenon include the nineteenth-century virtuoso Liszt, who played the piano with a tenderness and violence that contemporaries read in gendered, sexualized terms, and B. B. King, who famously named his guitar Lucille and spoke about the instrument like a romantic partner with its own desires.[84] The relationship between player and instrument not only has a history of being sexualized but is also revealing of gendered assumptions about agency. In *The Joy of Sex*, first published in 1972, Alex Comfort described a model of sex he called the solo, which is "when one partner is the player and the other the instrument.... The player doesn't lose control.... The instrument does lose control." Comfort argued for player and instrument positions being equally available to each gender but acknowledged that "the antique idea of the woman as passive and the man as performer used to ensure that he would show off playing solos on her, and early marriage manuals perpetuated this idea."[85]

Though an antique idea, the femininized instrument for masculine performance retained cultural currency into the twenty-first century. The personifications of keyboard instruments discussed in the previous chapter

FIGURE 5.2. Calvin Harris's organicism in the music video for "Summer" (2014).

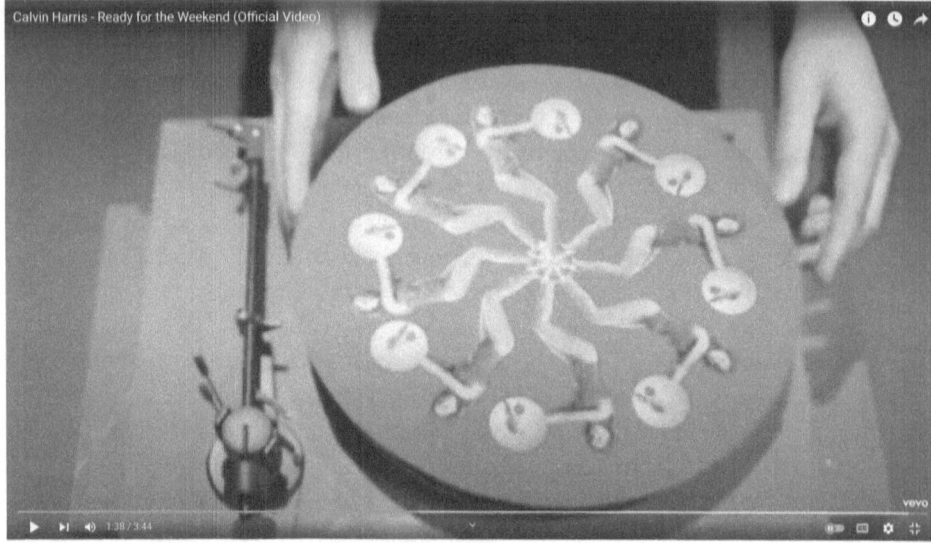

FIGURE 5.3. Woman-as-musical-instrument played by Calvin Harris in the music video for "Ready for the Weekend" (2009).

suggest how non-inevitable an *active player–passive instrument* configuration is and how significant stereotyped gender roles have been for instantiating this particular relationship. A version of the woman-as-musical-instrument shows up in Harris's oeuvre in the music video for "Ready for the Weekend" (2010): women wearing swimsuits become the surface of the LP on Harris's turntable (fig. 5.3). As Harris then self-effacingly sings "pushing knobs, pushing faders / But I, don't know what they do," he seems to admit novice status in both musical and sexual domains. Harris left behind this novice image when he "quit" singing and established himself as a producer skilled at manipulating others' voices. "This Is What You Came For" became part of Harris's DJ set; in performance he stands behind the decks controlling the playback of the song, Rihanna's vocal included. In this setting, Harris can be perceived "playing" Rihanna's voice like an instrument, particularly when the last word of the line "she's looking at you" is extended into a series of "ooo's," the clipped attacks of which make audible Harris's hand in manipulating the vocal.

The idea that female voices are the sound of passive instruments reaches a logical extreme in the vocal aesthetics of Snow, who told an interviewer in 2015, "Our current infatuation with manipulating vocals to sound like a computer and vice versa—which is 100% integral to everything I'm doing now, Terror Jr, Kiiara, etc.—is the first tiny step toward the computer doing all of the singing."[86] Snow went on to say that EDM festival audiences do not care about live singing and would be perfectly satisfied with a robot (a point that seems intended not to propose putting singing robots on stage, but rather to dramatize festival-goers' complete indifference to singers' embodiment or liveness). Together with Snow's view that singing "like a girl" is the "only way it works," these statements suggest that Snow produces female vocal-centric music while understanding women singers as mere sound sources, vocal instruments fully replaceable by computers or robots (once the sound synthesis is good enough). His own role as producer, on the other hand, shows no signs of future replacement by technology—suggesting that his work is that of a creative mind that cannot be digitally or mechanically simulated.

For her part, Kiiara seems to regard the melodic hook made from her chopped and pitched vocal as "the computer doing the singing," in contrast to the verses and refrain she sings herself. Though she lip-synchs the latter in the music video, she does not lip-synch the chopped hook; nor does she sing it in live performances, rather allowing the prerecorded samples to play. Listeners, however, generally do not draw the same distinction between Kiiara and computer as musical performer, as Kiiara learned from the phenomenon of people trying to figure out what she was saying. As she explained in an interview, she did not anticipate such efforts because she always knew "it's a sample. That's all it is."[87]

That listeners extended their expectations for meaning from Kiiara's singing to the sample-based hook owes much to Snow's production. As Kai Arne Hansen has observed, the way a voice sits in a mix can imply its relationship to a body, an aspect of production he terms "sonic staging" of the body.[88] In "Gold," the chopped hook occupies the same foreground place in the mix as Kiiara's singing on the verses and refrain, with the same sonic quality (EQ, reverb, etc.). The transitions between Kiiara's singing and the chopped hook are seamless and without overlap, encouraging the listener to imagine one and the same body singing continuously. This body thus seems to morph between human intelligence and computer intelligence, the latter sounding nonsensical but also—thanks to the articulate speech sounds, and as Kiiara's would-be decoders attest—as if it is trying to communicate, raising the possibility that it is our limited intelligence rather than the computer's that prevents us from understanding. One way to read the resulting polymorphous body is as a cyborg—as Brennan Carley did in *Spin* magazine, describing Kiiara as "half-woman, half-machine, equal parts human and power tool."[89] Snow's comments, however, combined with Kiiara's decisions not to performatively embody the "machine" half in live settings or the music video, suggest hearing "Gold" as giving sonic expression to Snow's idea of being on a path toward "the computer doing all of the singing."

"I LOVE WAILING"

MNEK's music aesthetics and creative process offer an alternative to the set of mutually supportive beliefs about voice and gendered bodies identified thus far—including the rigid gender differentiation of vocal range, differential availability to electronic manipulation, female bodies as passive instruments for male performance, and the replaceability of singers by computers. MNEK is a singer as well as producer, a dual role he performed on the track "Never Forget You," which he produced for Larsson's album *So Good*. MNEK has not expressed the same ambivalence as Harris about being a vocalist. On the contrary, unlike Harris and Snow, MNEK speaks of vocal abilities being common across genders. He explains his own development as a singer in terms of emulating Black women singers, but their gender difference or racial likeness to his own identities goes unremarked. Of his own singing, he has said, "I love being able to do that.... I love wailing, and I listen to Mariah [Carey], so I have to." Unlike Larsson, who naturalized vocal talent in contrast to learned production skills, MNEK acknowledges the study involved in learning how to sing: "I grew up with so much of the pop diva thing, like Whitney [Houston] was one of my most favorite artists. I love Mariah and I love Beyoncé. I listen to all the voices and I'd always try to emulate their riffs."[90]

MNEK's gender-fluid attitude toward voices comes with a similarly contrasting musical practice. "Never Forget You" features no seamless transitions between the singer's seemingly live, embodied performance and their audibly manipulated "pitched and chopped" vocal—no lead vocal lines assembled by chopping and pitching, as heard on both "Gold" and "This Is What You Came For." MNEK instead reserves highly processed vocals for background textures. Their most extensive use in "Never Forget You" occurs on the first post-chorus: a looped "aaaahh" from Larsson occupies the higher frequency range; the end of each four-bar phrase is punctuated with various vocal fragments, including the last phrase of the chorus—"till the day I die"—pitch-shifted to the bass register with an extra stutter, "till till till the day I die." And on the second eight-bar phrase, a highly processed wordless vocal sample (of indeterminate source—it could be Larsson's or MNEK's voice and indeed could be taken for a synthesizer) is used to create a stuttering accompaniment line. The processed vocal samples thus contribute to creating a layered collage of sounds, in distinct contrast to the lead vocal lines that are foregrounded in the mix and comparatively unprocessed.

Rather than present women's voices as interchangeable with computers or more available to electronic manipulation, "Never Forget You" presents women's and men's voices as largely overlapping in their capabilities. When Larsson and MNEK sing in harmony, it is sometimes Larsson supporting MNEK, her voice at a lower level in the mix, sometimes vice versa. They also do much of their singing in the same range. MNEK sings the second verse of the song in the same octave that Larsson sings the first. On the bridge, Larsson and MNEK trade phrases, repeating the same melody in the same octave in close succession. Both also perform wordless vocal runs at various points throughout the song. Just as he sampled and processed Larsson's vocals on the post-chorus of "Never Forget You," he does the same with his own voice in his solo work—as on the post-chorus of his 2016 single "At Night (I Think About You)," which layers a melody line of his highly processed vocal (to the point where it could be mistaken for a synthesizer) and relatively unprocessed, more rhythmic samples of his voice (audio 26). The differentiated treatment of female voices from his own voice evident in Harris's work in both range and types of processing is thus not evident in MNEK's.

In a behind-the-scenes video, MNEK demonstrated how he sang— holding his nose to alter the sound—and digitally processed his voice on "At Night (I Think About You)." As he explained, "That's my favorite thing about vocals. I love being able to just do whatever I want with any sound and manipulate it and tweak it and twist it up. It gives me the creative freedom that I always want when it comes to making music."[91] Rather than describing the computer as liberating the voice from the body, MNEK demonstrates the interaction of body- and computer-based vocal manipulations. There is thus a similarity between the singer's body and the computer, not because the latter can simulate and replace the former, but because both are highly flexible creative tools. MNEK's own body—as much as his vocal collaborator's—can be deployed as an instrument.

GENRES OF BEING POSTHUMAN

"If we are now posthuman, then what kind of posthumans will we be?" concluded Erik Davis in his 1999 *Village Voice* review of Hayles's *How We Became Posthuman*. Rather than "people not wanting to sound like people anymore," what we might hear in Richards's three songs are people differently negotiating how to be people in a posthuman world. Some people want to sound like women or like men, who have differentiated relationships to bodies, agency, and technology. In "This Is What You Came For" and "Gold," men perform their personhood through mastery and control of instruments—both computers and female voices. Women perform their personhood through their singing voice and their selection of collaborators. "Never Forget You," by contrast, lessens the difference gender makes, as well as the difference between body and technology—not such that a computer can offer freedom from biology, but because the body is already a creative tool.

While these songs can suggest experiences of being in the world ca. 2016, it remains for listeners to place them in history. The resources of cyborg and posthumanist theories offer divergent perspectives on such history. For Haraway, the cyborg is something we all are "by the late twentieth century." It represents a break with the world order of industrial capitalism and entry into that of an information system; it is disruptive of prior dualist hierarchies of all sorts.[92] Clark, by contrast, argues that we are "natural-born cyborgs" and that it is our cyborg nature that distinguishes humans from other species. Historically, humans have been cyborgs since at least the advent of speech (an early means of offloading cognitive work onto the world); and what we are witnessing now is an intensification of our species' proclivity for cognitive networking rather than something of a different order.[93] For Wynter, the advent of language is the very thing that necessitates thinking about the human species in terms of "genres of being human": the human's is a uniquely *bios-mythoi* existence, constituted not just biologically but storytellingly. "As humans," Wynter writes, "*we cannot/do not preexist our cosmogonies*, our representations of our origins—even though it is we ourselves who invent those cosmogonies and then retroactively project them onto a past."[94]

When the task is to debunk human-versus-robot criticisms of technology in music, Clark's vision offers an origin story that transmutes the threatening Other into a familiar part of the self, and does so while leaving other dimensions of social order largely undisturbed. The idea that we are and always have been cyborgs is a corollary to the idea that technology is part of being human. These ideas have a radical ring when directed against technophobic commitments to a "pure human" and do important work by opening space to appreciate artifice and hybridity; but in music criticism, they have also functioned as what Eshun termed "futureshock absorbers." That is, by assimilating the new to the familiar, they neutralize the potential for revolutionary reconfiguration.

To ask "*whose* body materializes" under the sign of the human, android, cyborg, or posthuman "limits the kinds of claims and interpretive leaps a writer can make."⁹⁵ It curtails the ability to make global statements about how people sound, how humans relate to technology, or how we became posthuman. But it also has the potential to generate new knowledge by helping bring into focus the persistence of long-standing structures in new guise or the emergence of new configurations. The sound of chopped and pitched vocals in ca. 2016 pop music contains the makings of both persistence and emergence—of listening toward how it sounds human after all or toward how music never sounds human alone.

CODA

Learning Machines

Lightning strikes from a purple sky on the 2009 album cover. The album title appears in white letters: *From Darkness, Light*. Above the title, a name is written in black: Emily Howell (fig. 6.1; audio 27).

Nothing about the album cover suggests that Emily Howell is a computer program designed by Cope. Like Cope's game asking listeners to "recogniz[e] human-composed music as distinct from machine-composed music," the album cover sets listeners up to perceive one thing (human-composed music) and then learn it's another (machine-composed music). In other words, it's a setup for a world-destabilizing revelation, setting in train questions like: If Emily Howell is the composer, does that mean computer programs are capable of creativity and self-expression? If Emily Howell can make human-sounding music, does it mean machines can do other human things—like feel, think, will? Does it mean they are on the doorstep of full autonomy and replacing humans?

As also with Cope's guessing game, such lines of questioning are products of the either-or framing of human and machine. In this case, they are additionally encouraged by the anthropomorphic naming of the computer program, which projects human status onto the machine. In fact, Cope has encouraged imagining Emily Howell as humanlike not only in compositional creativity but also in a more comprehensive kind of musical personhood (albeit in gendered terms that fall into the woman-as-instrument pattern discussed in the previous chapter). Thus, in the *From Darkness, Light* liner notes, Cope describes his role as being "to provoke her to compose good music, and once composed, act as her impresario—her agent—to obtain performances, recordings, and so on." Eleanor Sandry has noted that Cope slips between describing Emily Howell and its predecessor program EMI "as creative individuals in their own right" and as dialogic partners in an interactive process through which he actively shapes the musical results; but because Cope's emphasis falls on casting

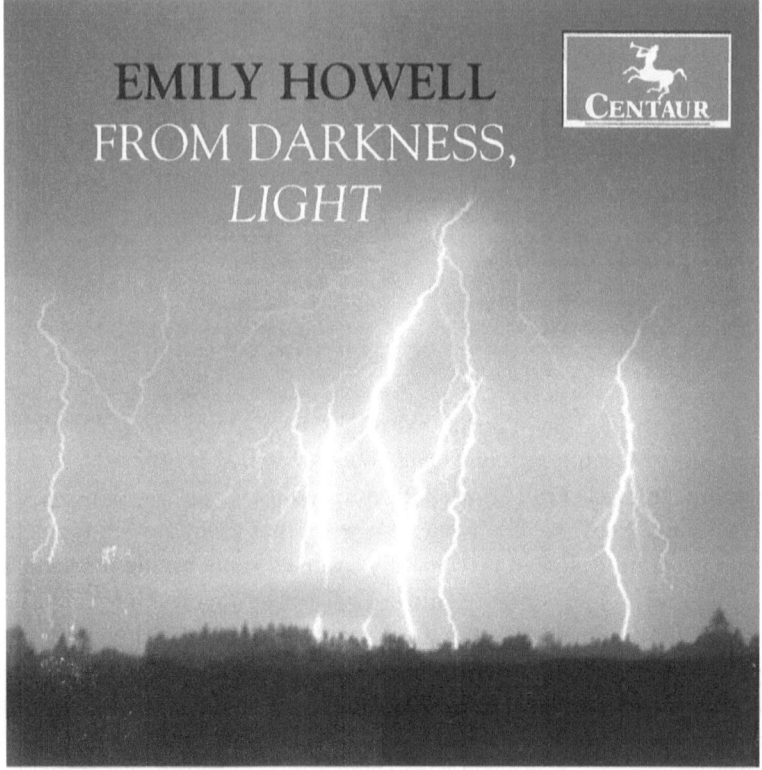

FIGURE 6.1. Album cover to *Emily Howell: From Darkness, Light* (Centaur Records, 2010), positioning Emily Howell as the artist; Emily Howell is the name of a computer program by composer David Cope. By permission of Centaur Records. Image courtesy of Lewis Music Library, Massachusetts Institute of Technology.

Emily Howell "as humanlike in itself, not just as mimicking human composers with its own type of creative processes," the effect is to obscure "the otherness of the machine."[1]

If full autonomy represents one pole of AI in the socio-musical imaginary, at the other is AI as "merely" a tool. Douglas Eck of Google Magenta, for instance, rejects framing AI as autonomous and instead foregrounds its dependence on humans. "We're not chasing the idea that we can make a human-feeling music without humans," he explained in 2017.[2] Instead, in the Google Magenta vision, human autonomy will remain unchallenged while human creativity is enhanced. As Eck told Arthur Miller, the goal with developing AI tools is to "increase our creativity as people [with] a cool piece of technology to work with."[3] The idea of enhancing human creativity by means of technology is not unique to AI; it is, rather, a common way in which music technology companies frame their products in the era of computers.[4] The music production software Reason, for example, was marketed with the slogan "This machine amplifies creativity" (fig. 6.2). Likening AI tools to earlier "unintelligent" technologies both slots them into an existing marketing logic and casts them as passive instruments, uncreative in themselves. Sony's Flow Machines, a suite of tools with capabilities such as generating melodies or harmonic accompaniments in particular styles, carries the tagline "Augmenting creativity with AI."[5] François Pachet, who led the development of Flow Machines at Sony Computer Science Laboratory in Paris, likens musical AI to digital synthesizers: "That was just a tool, and this is exactly the same thing. The tool does not have any idea about what it creates, and this is just a new generation of tools."[6]

Somewhere between the tool that passively amplifies and the robot that takes over lie emergent properties and relational possibilities that call for a richer conceptual vocabulary, a more expansive imaginary for what so-called AI is and does. While marketing copy and sensational headlines have remained directed toward the poles, musicians have been actively exploring the in-between space, developing "a new range of metaphor... a new set of analogies" (to recall Oram's words from 1972). George Lewis, for example, named the improvising computer music system he first programmed in the 1980s *Voyager* and describes a performance with Voyager

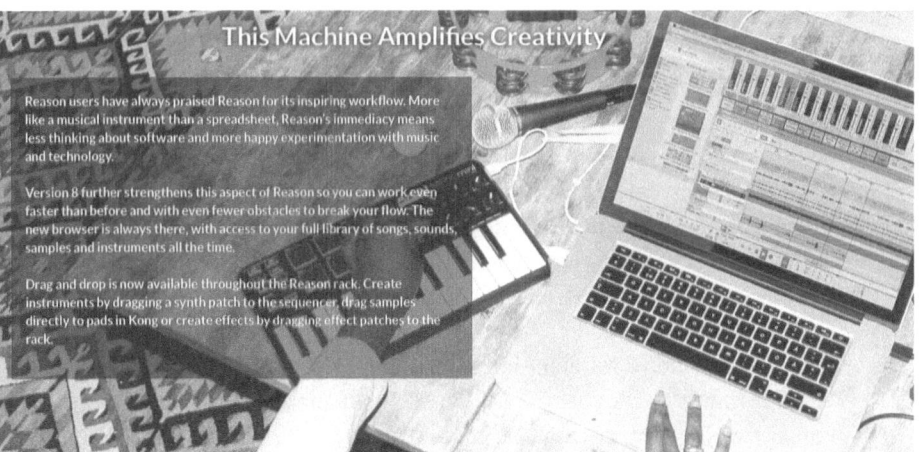

FIGURE 6.2. "This Machine Amplifies Creativity," said in advertising copy for the music production software Reason 8 on the Propellerhead Software (now known as Reason Studios) website. Software released September 2014; screenshot taken by the author February 16, 2015.

as "multiple parallel streams of music generation, emanating from both the computers and the humans—a nonhierarchical, improvisational, subject-subject model of discourse."⁷ In a discussion following a *Voyager* performance with pianist Vijay Iyer in 2012, Iyer drew an analogy between the system and a child, not to imply that it was poised to mature into human adulthood but rather to describe a quality of relationship based on its specific capacities: "The fact is we want our kids to be independent, but sometimes we know that they need help doing that; so there was a certain sense I got from George just before we went onstage that there was what he wanted it to do and what it might do anyway."⁸ Through such practice and reflection, musicians offer both experiences of and metaphors for the creative capacities and agential scope of nonhuman participants in their music making. In other words, they offer sounds and conceptual resources of heightened importance for shaping the currently shifting landscape of people and machines.

INTRODUCING NEURAL NETS

I first encountered the term *neural nets* in 2013, in a talk by sound artist Laetitia Sonami. As Sonami walked us through different phases of her creative practice using different interfaces, she mentioned neural nets almost in passing. The talk concluded with a brief performance in which Sonami dangled a "bird" from a string—a bird, she told us, she had "trained" for this performance (fig. 6.3; audio 28). "We'll see what the bird has to say," she remarked before beginning. Swinging the bird back and forth and lifting it up and down, she altered its position in relation to magnetic coils on the floor, thereby increasing or decreasing the range (in pitch and variety) of clicks and chirps. This, I was given to understand, was neural nets in action. I was left with confusion and curiosity, as I tried to make sense of what I had seen and heard. Did neural nets have some biological materiality? Were they fibrous nets? What did it mean that she had trained the bird to make these sounds—and that, at the same time, it was uncertain what the bird would "say" on this particular performance?⁹

Little did I realize at that 2013 event, neural nets (or neural networks) are basic to the field of artificial intelligence, enabling various kinds of machine learning. In the years since Sonami's talk, neural networks and machine learning have become far more ubiquitous in both discourse and practice. Google's DeepDream, released in 2015, was a significant conversation starter about neural networks, as well as a milestone in technical development and accessibility. Designed for the mundane task of classifying images (e.g., recognizing that an image contains a cat), DeepDream's

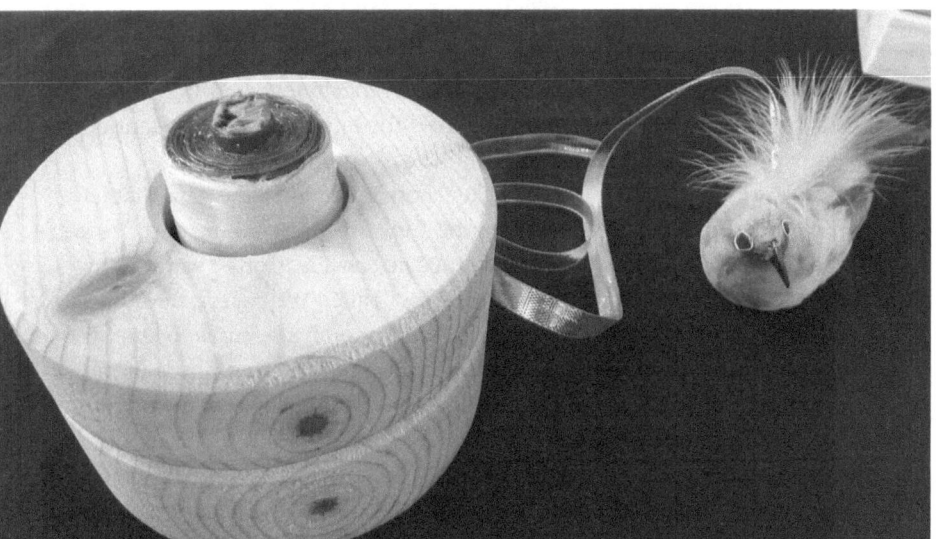

FIGURE 6.3. The author's introduction to neural nets takes the form of a bird: Laetitia Sonami, *Birds Without Feet Cannot Fly*, 2013. © Laetitia Sonami, reproduced by permission.

convolutional neural network could—through a kind of reverse process—modify images to bring out features that matched what it "recognized." With enough iterations of this process, the results were surreal and trippy images, with features like eyes multiplying in unlikely places.

DeepDream seemed to be probing computers' unconscious, naturalizing computer-brain analogies and helping establish neural nets as a dormant metaphor, a term native to computing. It was by explicit analogy to neurons in the brain, however, that the principles of neural nets were first formulated. The term *neuron* was coined in the late nineteenth century to describe a conception of nerve cells as bounded, independent entities (in contrast to conceptualizing the nervous system as a continuous net of nerve fibers).[10] In the early twentieth century, physiologists observing the transmission of electrical impulses between neurons developed an "all-or-none law," theorizing that each nerve cell has a threshold that an electrical impulse must exceed in order to activate, or excite, the cell. Regardless of whether the impulse barely exceeds the threshold or exceeds it by a lot, the excitation of the neuron is the same—hence the "all-or-none" description and a binary, on/off conception of the neuron.[11] In 1944, the neurophysiologist Warren McCulloch and mathematician Walter Pitts proposed that due to this all-or-none behavior of neurons, it is possible to describe "neural events and relations among them... by means of propositional logic," propositions being either true or false.[12] As an approach to artificial intelligence—to the question of whether machines can think—neural networks went in and out of favor across the latter half of the twentieth century before taking center stage in the early twenty-first.[13]

Revisiting Sonami's work today, it is possible to fill in the gaps that left me wondering about nets and training—in some ways taking away from what I imagined in 2013 (goodbye, fibrous nets) while in other ways adding to it. Sonami used a machine learning tool called Wekinator, developed by Rebecca Fiebrink. Wekinator facilitates mapping gestures to sounds. Training involves creating a set of examples pairing the inputs of performer actions (such as holding the bird at a certain height) to the outputs of sound synthesis parameters (in the software Max/MSP). Through this supervised learning approach (supervised because the training data specifies the correct output for a given input), Wekinator builds a model that computes synthesis parameter values in response to the performer's actions. Sonami calls this model a "synthesis terrain"; shaped by the training data, the terrain also fills in synthesis parameters for actions not included in the training data. Asked whether she thinks about machine learning more as a tool or as independent intelligence, Sonami replies: "I wish I could say I approach ML as having independent intelligence or agency, but I essentially use it as a mapping technique and it is part of a system." While not sensing intelligence or agency in the machine learning component alone, however, Sonami does attribute such qualities to the system as a whole—which is to say, to her instrument. Describing the instrument as comprising the hardware, sound synthesis software, and machine learning used to map between them, Sonami remarks that it "does have agency and identity" and observes that it took years of working with this system "to get to a point when I started to understand 'what the instrument wants.'" Sonami has worked toward her performances becoming "more of an 'exchange' between the instrument and me, the performer. Not just forcing my intentions onto it, but letting it inform the composition and performance."[14] Sonami uses machine learning to obtain unpredictability within parameters, creating a "synthesis terrain" that she navigates through gesture and in which she discovers sounds she did not explicitly program. She describes performance as a practice of listening, learning, and responding to the instrument's behaviors—to "what the instrument wants"—figuring the instrument more as a collaborator than as an extension of herself.

Having desires but not sentience, agency but not autonomy—machine learning contributes to these properties of the instrument, which make it a collaborative partner without making it a human person. These properties are not unique to machine learning, however; musicians have formed similar relationships to instruments such as guitars, synthesizers, and (as we saw in chap. 4) pianos. To the degree there is a *same thing* quality to machine learning tools, it is not in their being just another tool for the artist's expression but in their being a new generation of musical collaborator—which is not to say that all is fine in innovation, but that it really matters what these tools can do in our relationships with them.

"A NEW SET OF ANALOGIES"

Since Sonami's talk, new types of neural networks have been developed, and musicians have continued to offer opportunities to experience and reflect on what they are and what they can do. Consider the album *Coditany of Timeness*, released by the duo Dadabots in 2017. As their band name suggests, Dadabots are interested in both automation and making art that is satirical, irreverent, and subversive of art itself. The year 2016 was early days for generating audio with neural networks—for neural synthesis—and Dadabots were among the first to apply the techniques to generating music. They generated *Coditany of Timeness* using a recurrent neural network based on SampleRNN, which was originally developed for text-to-speech synthesis. They trained the network on the album *Diotima* by Krallice; as they noted on the album's website, *Diotima* was "the original album we trained our bot with."[15] Unlike Sonami's bird, which conjures life and organic processes, *bot* connotes computer automation. The image of a bot trained to emulate an existing band sets up the conditions for comparative listening, querying how human the machine-generated music sounds. Anthony Fantano, music critic of the YouTube channel The Needle Drop, found the results surprisingly convincing. "That's really weird," he said in reaction to the album. "It makes me smile because I'm actually kind of amazed by it.... The aesthetic is there.... It sounds like I'm honestly listening to a live demo of Krallice.... It sounds like I'm actually hearing regular instrumentation, like regular black metal, atmospheric black metal instrumentation."[16] Dadabots indeed found that their SampleRNN approach more successfully emulates metal and punk than other genres like electronic music, "perhaps because the strange artifacts of neural synthesis (noise, chaos, grotesque mutations of voice) are aesthetically pleasing in these styles." The fast tempos and loose ensemble coordination additionally "translate well to SampleRNN's rhythmic distortions."[17]

In 2019, an interviewer expressed surprise to Dadabots member Zach Zukowski about their success with death metal because they would have thought this music was "too complicated to replicate with an algorithm," in contrast to "simple, straightforward pop music," which they would have expected to be easier to replicate. The encounter with Dadabot's music thus challenged the interviewer's preconceptions about music and machines, revealing an image of pop music as already algorithmic in quality (simple, formulaic) and of death metal as unruly, too complex to be expressed algorithmically. Zukowski replied that the interviewer would be correct if they were dealing with symbolic representations of the music that reduce it to fundamental pitches and rhythms rather than with audio as the training data. But it is the very idea of training data in connection with machines that gets at the source of surprise; the interviewer brings to the conversation an image of the music-generating machine as something deterministic, requiring explicit musical rules to follow, rather than as something that finds patterns in data and uses what it thus "learns" to generate new data. Such a shift in understanding the machine helps make sense of what Zukowski explains: the fast speeds of death metal—features like rapid guitar playing and blast beats in the drums—mean that the network can "learn more of the structure in a shorter amount of time."[18]

Since 2017, Dadabots have continued to experiment with the possibilities of neural synthesis and to play with the fears surrounding AI and automation. In 2020, they "trolled" the Eurovision AI song contest by submitting a death metal track about robot takeover. "Vote for this song, your AI overlords command you," reads the screen of the music video before the song begins with a toneless voice rasping, "Human extinction is the only way." In 2021, Dadabots released an album titled *Can't Play Instruments Vol. 1*, which narrates the duo's loss of musical skill and well-being. The album cover art illustrates the second track, "We Generated this Album in Our Sleep" (audio 29); a computer screen shows the word GENERATING while the duo slumps unconscious (fig. 6.4). "Our Musicianship has Completely Atrophied," the third track is titled. "We Haven't Listened to This Song Yet" (audio 30) reads the title of track ten, and the unusually abrupt ending to the track, which unlike others stops mid–musical thought, suggests the band indeed put minimal curatorial effort into the song (or else chose it precisely for such effect).

Behind the artistic front of humanless creation, however, Dadabots offer a very different story about what they do. For the 2020 AI song contest, for example, they "curated" their favorite lines from "over 10k lines of very dismal, brutal, and gory lyrics" generated by a GPT2 neural net trained on death metal lyrics.[19] Similarly, their 2017 albums constituted curated selections from hours of generated music. CJ Carr informally estimates that they considered only about 5 percent of their AI-generated tracks at that point "usable." The 2019 release of "RELENTLESS DOPPELGANGER"—a

FIGURE 6.4. Album art for Dadabots, *Can't Play Instruments Vol. 1* (Dadabots, 2021). Artwork by Dadabots, reproduced by permission.

"neural network generating technical death metal, via livestream 24/7 to infinity" on YouTube—marked a turning point for the duo; they had learned how to craft the training data and the network so that "everything it makes sounds great" (making possible a continuous stream of newly generated music). Carr has described this as getting better at "using the neural nets as an instrument"—a stark contrast to the "can't play instruments" image.[20]

In 2019, Dadabots collaborated with beatbox artist Reeps One on a project aimed at "training a machine to emulate his voice."[21] A documentary portrays the process: Reeps One provided audio recordings of himself performing to Dadabots, and on-screen Carr plays audio generated by the neural network at different stages of training. At first, it generates noise. Gradually, patterns learned from Reeps's audio recordings begin to emerge. "I can already hear it growing," he comments. "It does make me think of like an embryo." The embryo image is suggestive of both the unformed quality of the audio and the potential for development sensed within it: the promise that from this noise, order, beauty, something more like oneself will emerge. Responding to a later stage of training, Reeps observes, "It's obviously still not me but immediately the [aspirates *ha hoo ha*], my harmonics, the shape of my mouth I can start hearing in the audio that it's generating."[22] Ultimately, Reeps performs a duet with his emulated voice, dubbing it his "AI 'Second Self.'"

Vocal modeling and developmental metaphors also feature in the work of Herndon and Mat Dryhurst. In 2018, they announced their "AI baby" named Spawn and released its first track, titled "Godmother" (audio 31). In a press release accompanying "Godmother," Herndon introduced Spawn as "a nascent machine intelligence" and as a member of her musical ensemble.[23] With Herndon and Dryhurst as mother and father, Spawn also had electronic music producer Jlin as godmother and programmer Jules LaPlace as godfather. Much like Reeps heard an embryo in the early training of his voice emulation, music critics picked up the baby image to make sense of "Godmother." NPR music critic Lars Gotrich described "a raw, newborn quality to the track as it hums and sputters like a swarm of glitching bees, just trying to find its mother."[24]

While aesthetically and narratively powerful, the baby metaphor is also technically obfuscating. The effect of the track is quite different when imagined to be not the sound of a pre-articulate voice, an incipient intelligence beginning to emerge, but rather a model trained on certain sonic materials being applied to sounds of another kind. The track was made with a neural network trained on Herndon's voice and applied to a track by Jlin, resulting in audio style transfer.[25] For an idea of the musical structures that might have characterized Jlin's base track, consider "Expand," from her 2015 album *Dark Energy*, and the suggestively titled "Embryo," from her 2021 album of that name. The former (which *New York Times* music critic Jon Pareles picked as a best song of the year in 2015) was also a collaboration with Herndon and features a driving synth at the same rate as the noisy vocal that starts off "Godmother," supported by a similarly downbeat-marking low thud and other relatively sparse percussive parts (audio 32).[26] "Embryo" features a frenetic pace of sonic activity, as well as a gritty timbral quality ("It's fun to listen to her toying with corroded sounds, degraded outputs," wrote Jayson Greene of the album), that feels in the realm of "Godmother" (audio 33).[27] Dryhurst has explained the aesthetic interest of "Godmother" for its makers being rooted in how unexpectedly the vocal model behaved with Jlin's production and how far removed from Herndon's own vocal performances. As Dryhurst remarked, "The 'beatboxing' for example, which is the feature of this track, was something the networks dreamed up, presumably learning from the stops and starts in

Holly's speech. That's kind of the reason we decided to lead with this piece, as it was a clear example of the networks making a decision that Holly never would have made."[28] Herndon explains of Spawn:

> She likes transients.... She saw a snare, and thought, *That is a bit like this bit I recall from when Holly says "T"* and tried to reproduce the snare with a "T" sound. That, to us, is new. The result is somewhat clever, logical, and most importantly, unexpected. It surprised us. That's why when you listen to "Godmother," it sounds like beatboxing, which is a combination of singing and speech.... It wasn't something I told her specifically to do. I tried singing "Godmother," and I just can't. It's too fast. Spawn outperforms me.[29]

This account—emphasizing the unexpected nature of what Spawn produces, as well as its ability to perform in ways Herndon cannot—sets up quite different listening expectations than the introduction of Spawn as an AI baby. It brings Herndon's approach closer to Sonami's, exploring a "synthesis terrain" shaped by the model. Herndon also echoes Sonami's experience of learning what the instrument "wants." Regarding Spawn's response to transients—the brief, nontonal sounds at the beginning of a waveform, particularly pronounced with percussive sounds like drum hits and plosive consonants—Herndon describes how it shaped the training data they created: "Sometimes we found ourselves doing things to please her," making the kinds of sounds "that she really likes."[30] In such creative work, it is as much the musicians as the neural networks that are learning through the training process.

In contrast to Iyer's use of a child metaphor to foreground the partial independence of Lewis's *Voyager* system, Herndon has explained that the baby image is meant to describe the immature stage of the current technology: "Spawn is our AI baby. And when I say that, it's basically a metaphor for the experiments that we've been doing with music using neural networks. One of the reasons we chose a child or baby metaphor when talking about Spawn is that we really see it as kind of nascent, baby-like technology. It's at such an early stage that we're trying to imagine what if this community, what if our approach to raising her could really have some sort of impact on how she grows up."[31] The question of how Spawn "grows up" is thus the question of the future of artificial intelligence. In the "Godmother" press release, Herndon asked whether "permission-less mimicry [is] the logical end point" of tools like Spawn that can learn to emulate existing voices and styles or whether there is "a more beautiful, symbiotic path of machine/human collaboration, owing to the legacies of pioneers like George Lewis, that view these developments as an opportunity to reconsider who we are and dream up new ways of creating and organizing accordingly." In addition to Lewis's work with computer improvisation as a practice of interacting with machines dialogically, these ideas resonate with Bruno Latour's argument for "car[ing] for our technologies as we do our children" and Jason Edward Lewis, Noelani Arista, Archer Pechawis, and Suzanne Kite's reflections on "making kin with the machines." As the latter argue, the human-centered approaches often touted as the antidote to runaway technology are ill-equipped to grapple with the interconnectedness of humans and nonhumans, creating a need for "frameworks that conceive of our computational creations as kin and acknowledge our responsibility to find a place for them in our circle of relationships."[32]

Along with a relationship of care and responsibility, however, Herndon and Dryhurst's baby metaphor implied that "growing up" means becoming more independent, more a fully formed person—a trajectory for AI reinforced by the contexts into which Spawn was placed. When the single "Godmother" was released in 2018, Spawn was credited as a featured artist and appeared on Spotify as an artist on the track along with Herndon and Jlin.[33] This choice matched the construction of Spawn as a member of the musical ensemble and formally recognized this computational creation as kin with a place in the circle of relationships. However, it also projected authorial agency and commercial personhood onto Spawn, like Cope did with Emily Howell. Commercial personhood was claimed for a number of other musical AIs around this time, garnering media coverage that sensationalized the leap from AI-generated music to AI as conscious, autonomous, and on the brink of fully replacing humans.[34] On the 2019 album *PROTO*, "Godmother" was the only track to credit Spawn as an artist, an apparent holdover from its prior release.[35] By 2021, Herndon came to prefer discussing Spawn and successor projects in terms of machine learning rather than artificial intelligence in order to avoid "Skynet" associations—a reference to the 1984 film *Terminator*, in which the AI Skynet becomes conscious and autonomous and endeavors to wipe out humans.[36]

Other musicians working with new machine learning capacities around 2020 have likewise found the prevailing cultural templates for AI ill-suited to their experience and have developed alternatives to the doppelgangers and progeny metaphors that center human likeness. Composer-performer Jennifer Walshe is one such musician. In 2018, she collaborated with computational artist Memo Akten on a musical project entitled "Ultrachunk." Akten designed a neural network called GRANNMA (Granular Neural Music and Audio), for which Walshe supplied training data in the form of hours of her vocal improvisations. In performance, Walshe improvises with the system in real time, relating to it as a vocal partner. In keeping with the tropes around Dadabots's work, a program note describes the system as her "AI doppelganger" and suggests the effect is both "spellbinding and deeply alarming."[37] In interviews, however, Walshe projects more awe and delight than alarm when discussing her experience with the system, and describes thinking about AI not as artificial intelligence but as "a desire to witness something that's alien." Invoking Haraway's notion of "companion species," Walshe sidesteps the organism-machine dichotomy embedded in the cyborg and instead compares the experience of improvising with the network to "when you ride a horse, or when you watch a farmer with a sheepdog be able to herd sheep in, and you think that's something more, and it's something different than human intelligence, it's something special."[38] The AI interests of sound artist and technologist Hexorcismos (Moisés Horta), who released his first album of neural network–generated music in 2020, are also oriented toward nonhuman intelligence rather than how humanlike AI can become. Hexorcismos describes "approaching the Neural Network as an alchemical technology"—the album is called *Transfiguración*—for morphing together his own electronic music with the album *Templo Mayor—Musica Con Instrumentos Prehispánicos* (1982) by Antonio Zepeda. Hexorcismos thus centered style transfer as the main operation of his neural network, which he used not to create his digital double or offspring but rather to continue a musical tradition of bringing precolonial sonic histories into the present and future.[39]

Arca (Alejandra Ghersi) has proposed another alternative to the concept of AI—"intelligent sentience." In 2019, Arca partnered with artist Philippe Parreno to provide music for an installation at the Museum of Modern Art (MoMA) in New York City. Arca composed music to be processed by software made by Bronze, a company with a vision for "audio files of the future" being not static but "fluid, intelligent, and capable of responding to external input."[40] As she described the process of making the sound installation called *Echo*, she "gave musical language to Echo, a musical being birthed into the Museum of Modern Art's lobby, and then Echo began to speak of its own volition." *Echo* generates music based on Arca's initial sound materials combined with data collected from the environment, reflecting what is happening in the MoMA lobby. Parreno describes *Echo* as "an automaton fed by various inputs and data.... You could say that *Echo* is a sentient automaton that responds to its surroundings by receiving data that us humans don't really perceive, but a machine can."[41] In other words, he adds, it is an "automaton that feels." A machine that senses things humans do not adds a possibility not present in the questioning around birds, musical instruments, moving statues, and androids we saw in chapter 1, where to feel or not to feel drew a key contrast between human and machine. To be sure, the difference between how *Echo* and, say, a harpsichord interact with their environment is a material one. But the question of sensitivity is also, as we have seen, an imaginative and analogical one; by describing *Echo*'s processing of inputs (which include data on atmospheric pressure, sunlight, and wind direction) as feeling and its audio outputs as speaking, Parreno and Arca vest *Echo* with subjectivity. Rather than necessitating the extension of the analogy toward human identity, this subjectivity holds together for Parreno and Arca with *Echo*'s machineness, marked by its distinctive perceptual capacities. How to understand *Echo*'s creative agency, by contrast, is a point on which Parreno and Arca diverge. For Parreno, the work's name reflects its *inability* to speak of its own volition; it is "doomed" to react to its inputs in certain ways, to "bounce back what it feels."[42] One might conclude that such dependent responsivity epitomizes the opposite of being human; or one may recall Diderot's use of a harpsichord analogy to explain how human thought arises from feeling matter in strikingly similar terms.

In 2020, Arca again worked with Bronze, this time to produce an album of a hundred remixes of a track from her album *KiCk i* called "Riquiquí" (audio 34). In a press release, Arca noted that "up until now I had never allowed anyone to remix an Arca song," at once putting Bronze in the position of first person to remix her music and suggesting there was something about its nonhumanity that made her open to the collaboration. The idea of a hundred remixes resonates with a consistent theme in Arca's work: a conception of the self as both singular and multiple. Regarding her use of a range of vocal registers and expressions on "Riquiquí," Arca observes, "You can see this character almost shape-shift... but there was something of an essence that unified all of them, and it was just electric." Across the hundred remixes, musical elements emerge from the background into the foreground (like the humming and its ethereal double at the beginning of #25 and #47, drawn from around 2:00 in the original track; audio 35), subtracted (like the lead vocals, which are sometimes absent, as on #33 and #53; audio 36) and transformed (as in the introduction section of #77, where a descending melodic figure familiar from the original version sounds like it is generated anew by means of a granular synthesis technique; audio 37). All, however, preserve in some form the pulsing synthesizer and ricocheting percussion heard at that start of the original track (#86 being the exception that proves this rule). And all exist in the same sonic terrain, based on the percussive, electronic, and vocal sounds of the original track. By working with "Bronze's trained yet unpredictable musical AI," Arca described being able to experience novel versions of her own music: "For a composer such as myself it remains something truly new which I had never experienced before, a moment of unforgettable experience in virtue of the mystery and wonder Bronze makes possible."[43] Through a hundred AI remixes of "Riquiquí," Arca intimated the possibility of a "forever-mutating instance of the song.... A Prometheus flame."

The album art for "Riquiquí" is a QR code that is supposed to be an "instant gateway" to that "forever-mutating instance of the song" (fig. 6.5). Scanning the QR code made me keenly aware of the "intelligent sentience" of my phone, as I found the proper distance at which it could "see" and "read" the arrangement of black-and-white squares specially designed for machine sensing. It was through this activity that I noticed that each black square of this QR code, rather than the usual solid color, has a faintly visible portrait of Arca. These portraits make no difference to my phone's reading of the QR code; they transform the image for me, revealing its design for both human and machine vision, each able to do something with the image the other cannot. Once my phone brought me to the "Riquiquí" web page, however, I was met not with a song's infinite life but with silence; although a play button appeared, it made no music sound. Viewing the source code for the web page reveals text that states, "Your browser does not support the audio element"—it is May 2022, less than a year and a half after the album was released. A demonstration of the potential for endless generativity turned into a demonstration of the fragility of digital information flows, so interdependent with their computational and material infrastructures—with human-and-machine activities.

In an interview with Matt Moen for *Paper Magazine*, Arca discussed her conception of self as both unified and multiple, connecting the idea to songs like *KiCk i*'s first track "Nonbinary" (audio 38): "I'm asking for recognition that we have multiple selves without denying that there's a singular unit. [It's] the difference between the pronoun use of 'they' and 'it.'... I want to be seen as an ecosystem of minor self-states without being stripped of the dignity of being a whole."[44] Arca's request for recognition of such selfhood is both general to humanity and specific to herself as a trans woman. Transitioning in the public eye, Arca has sought to replace the idea of gender dysphoria (implying a mismatch between singular, binary gender identities) with gender euphoria, a pleasure in experiencing multiply gendered selves.[45] On album covers and in music videos, Arca makes extensive use of imagery of body modification through surgery and prosthetics, embracing such mutability and granting no special status to a "natural" organic body. The opening lyric of "Riquiquí"—"*rosa blanca de metal*" (white metal rose)—offers an image that likewise blends organism and machine in a form more unitary than hybrid. In this light, it is

FIGURE 6.5. *Top:* The album art for Arca, *Riquiquí; Bronze-Instances (1–100)* (XL Recordings, 2020), encodes a website address in the QR format designed for machine reading. *Bottom:* Using my smartphone to read the QR code and visit the website revealed a close-up of the QR code in which variations on Arca's portrait were more clearly visible in each dark square, as well as an audio player that was meant to play the album but did not work.

especially significant that Arca jettisons the term *artificial* from AI by referring instead to intelligent sentience. Whereas *artificial intelligence* reiterates a relation of emulation, of derivative and imperfect copy to the "real," "authentic" intelligence of humans, *intelligent sentience* encourages engaging with the particular qualities and behaviors of algorithmic systems on their own terms—opening up alternative ways of imagining their place in the world and potential futures.

These examples cohere into an era of music ca. 2017–2020 made with neural audio synthesis, marked by gritty sound quality, glitchy and granular aesthetics, and a sense of ontogenetic becoming towards human likeness or intelligences unknown. Meanwhile, machine learning tools have also been developed to "assist" musicians in a host of compositional and audio engineering tasks, intended—like Auto-Tune was—to save musicians time and money without introducing any audible difference.[46] At this juncture in the early 2020s, one of the uses for machine learning being pursued is *timbre transfer* (a more targeted audio style transfer). Differential digital signal processing (DDSP), introduced by Google researchers in 2020, has made it possible—as the researchers illustrate with an audio example—to record one's singing voice and then transform it into the sound of a violin.[47] The result is the same melody in a violin-like timbre along with features very unlike any violin playing I've ever heard (and especially unlike the performance of Bach's Violin Partita No. 1, from which the training data for the model was drawn). For instance, the vibrato and slides come from the vocal performance, departing stylistically and technically from the sound of a violinist producing such pitch inflections with their fingers. There are also unusually loud and frequent sounds reminiscent of fingers hitting the fingerboard, likely the model's translation of the singer's consonants. The sound is thus more than a hybrid of opposites; we can hear in it the voice, the violin, and the machine learning model.

Among the efforts to make timbre transfer accessible to musicians is a digital audio workstation (DAW) developed by Yotam Mann and Chris Deaner under the name Never Before Heard Sounds. Describing a desire orienting their work, Deaner noted: "I can't play a trumpet very well... but I will often hear that timbre, I will often hear that sound and I will want it when I'm creating something. So if you go to this DAW that we're creating right now, one of the few things you can do is select a model to transform your input into. So then out will come, could be a trumpet."[48] This remarkable echo of Oram's description of "wanting to make some sort of trumpety sound" in 1972 (see chap. 3) throws into relief a significant difference: the dream here is focused on the singing voice rather than the writing hand as the musician's point of contact with the machine. This shift suggests the potential to collapse the gendered division of labor between singing and producing discussed in chapter 5. But it also provides occasion to ask *what of the hand?*—not out of an idea that there can be no human expressivity without it, but out of concern that the falling away of the hand is symptomatic of the logic of AI (with a through line back to the becoming android of Vaucanson's flute player), which regards anything performable by machines as no longer a matter of human intelligence and imagines hands to be always already mechanizable rather than sites of embodied knowledge.[49]

Listening to these musicians' music, along with what they say about their work, expands the possibilities for thinking about what *AI* is and does, for experiencing our relationships to machines, and for shaping our futures with them. Moving beyond either-or framings of human and machine remains, as of the early 2020s, an uphill journey. Consider the *New York Times* coverage of the second international AI Song Contest, held in 2021; the contest framed AI as a creative and collaborative tool for musicians, and the journalist noted that musicians discussed "how they worked with the technology both as a tool and as a collaborator with its own creative agency." And yet, the story ran with the headline "Robots Can Make Music, but Can They Sing?," repeating the boundary line drawing we saw emerge around musical androids after 1750 (Quantz's *They may astonish, but never move you*). The journalist also noted that "many of the results wouldn't sound out of place on a playlist among wholly human-made songs," as if songs made without AI did not likewise involve nonhuman contributors.[50]

The history of human-machine relations traced in this book supplies material for thinking outside the narrow, well-trodden path of incremental technological encroachment upon the uniquely human. Through histories of musical automata pleasurably imagined to be sensible, string instruments considered machines, the voice as a hybrid string-wind instrument, bells sounding harmonious rather than nonhuman, philosophers being like harpsichords, machines needing to be humanized, humans being made up of tuned circuits, pianos being like people, and people sounding posthuman, we have seen *human* and *machine* continually renegotiated. I do not propose that we find a solution, a forgotten key to how we should think about humans and machines in this history. Rather, I propose that its various musical human-machine configurations can have a helpfully dispersive, centrifugal effect on our thinking, experiences, and decisions. By weakening the grip of *But can a machine... ?* reactions, we make way for other questions, other interests in what we and our machines can do, and chances to build what we learn into what we become.

Acknowledgments

The idea for this book grew out of discussions in the classroom. Thank you to the many students who have worked through questions about music and technology with me and who turned the thing in need of doing into figuring out how human and machine became musical opposites, as well as what reasons and resources there are for thinking otherwise. Preparing the way for these conversations were, in turn, my own student and postdoctoral fellow days; thank you to the teachers and mentors to whose model I continue to aspire.

For financial support of extended time to research and write, thanks to my institutional home, Northeastern University, and to the Institute for Advanced Study in Princeton, New Jersey (Edward T. Cohen membership in Music Studies, 2019–2020). Travel to the Daphne Oram Collection at Goldsmiths, University of London, was made possible by a summer research travel grant from Northeastern's College of Arts, Media and Design (CAMD). CAMD also funded student research assistants who brought crucial insight to the state of music AI and the sound of ca. 2016 popular music: Sadaf Khansalar and Chris DiPierro in the early exploratory research phase and Grant Foskett in the final stretch.

Invitations from a number of people, and the conversations to which they led, proved formative for the project: Jason Farago to write on the long history of music technology for *Even Magazine*; Roger Moseley and Annette Richards to speak at the Westfield Center "Keyboard Networks" conference and to elaborate on pianos going to the dump for *Keyboard Perspectives*; Ann Powers to submit a paper for Pop Con; Emily Dolan and Emily MacGregor to set readings for a study day on Musical Thought and the Scientific Imagination; Karen Desmond to speak at Brandeis University; Tara Rodgers to moderate her performance and Q&A at the Women, Feminists, and Music conference; Matthew Head to think about metamorphosis in eighteenth-century music; Elizabeth Margulis to write about humanistic and scientific conceptualizations of music for *Music Research Annual*; Gabrielle Ferrari on behalf of the Historical Musicology and Theory areas to speak at Columbia University; Edward Halley Barnet and David Byrd to speak at Hamilton College; and Eamonn Bell, Ezra Teboul, and Brian Miller to share work in progress with the Consortium for History of Science, Technology and Medicine working group on Sound and Technology. Thanks to each of you for these opportunities and to all involved for the fruitful discussions that helped make this book better.

For fielding questions that came up along the research way, thank you to Laura Crook Brisson, CJ Carr, Seth Cluett, Anthony De Ritis, Holly Herndon, Francesca Inglese, George Lewis, Nick Patterson, Jasia Reichardt, Tom Richards, Caroline Rusterholz, Scott Sanders, Victor E. Sachse, and Ben Schneiderman. Leslie Ruthven facilitated access to the Oram collection, making it a smooth and enjoyable archival experience. Additional encouragement for the project came from conversations at conferences where I presented research in various stages of development. These included a session convened by Olivia Bloechl and Melanie Lowe at the American Society for Eighteenth-Century Studies; a session organized by Kyle Devine and Patrick Valiquet at the Music and Philosophy Study Group of the Royal Musical Association; a session organized by Etha Williams at the American Musicological Society; the Recursions: Music and Cybernetics in Historical Perspective conference organized by Patrick Valiquet and Christopher Hayworth; and the Society for Literature, Science, and the Arts conference on "Reading Minds."

Through generous reading of chapter drafts, a small village has helped hone this book's ideas and writing. For feedback in early stages, thank you to Amy Coddington, Lauren Flood, Robin James, Nick Seaver, and Marie Thompson. At IAS, I benefitted from the collective wisdom of the History of Science seminar led by Myles Jackson and the Early Modern Europe seminar led by Francesca Trivellato. Likewise, deep appreciation to my fellow fellows in the Northeastern University Humanities Center Re-imagining/Re-forming group, led by Patricia Williams. For crucial encouragement, suggestions, and corrections toward the end, thank you to Psyche Loui and the MIND lab, Adeline Mueller, Chris Parsons, Hilary Poriss, and Griffith Rollefson. For heroically reading an entire draft manuscript and zeroing in on its strengths and weaknesses, a special thank you to Ellen Lockhart, Thomas Patteson, and Nick Seaver. I am also grateful to the anonymous readers of the initial book proposal and full manuscript, whose constructive feedback provided guiding lights, and to Suzanne Bratt for above-and-beyond indexing. For their steadfast support of the project from idea to final form, thank you to Nicholas Mathew and Marta Tonegutti. And to Adrian Freed, Clara Latham, Andrew Mall, Rebekah Moore, Mary Ann Smart, Kira Thurman, and Rachel Unger, thank you for sharing food for thought and actual food—I'm so grateful for the time we've spent together in the same place.

Portions of this book have appeared in earlier versions in prior publications. These are: "Metamorphosis and the Taxonomy of Musical Instruments," *Journal of Musicological Research* 40, no. 3 (2021): 279–88; "Daphne Oram, Cyberneticist?," *Resonance: The Journal of Sound and Culture* 2, no. 4 (2021): 503–22; and "Piano Death and Life," *Keyboard Perspectives* 10 (2017): 1–18.

Lastly, for making any of the preceding possible, thank you to my husband, Conor, who does the truly important work of making music. I love you, and Noah, Blaise, and Artemis are so lucky that you're their dad.

Notes

INTRODUCTION

1. David Cope, *Virtual Music: Computer Synthesis of Musical Style* (Cambridge, MA: MIT Press, 2001), 13.
2. Douglas Hofstadter, "Staring Emmy Straight in the Eye—And Doing My Best Not to Flinch," in Cope, *Virtual Music*, 38.
3. Hofstadter, "Staring Emmy Straight in the Eye," 80.
4. "Musical DNA," Radiolab, September 24, 2007, https://www.wnycstudios.org/podcasts/radiolab/segments/91515-musical-dna.
5. Cope, *Virtual Music*, 32. Cope also discusses how his computer-generated music challenges assumptions about the nature of musical creativity and about what goes on in musical listening.
6. For an examination of Chopin's music in relation to "the keyboard as a digital interface that can act as grid, filter, and analogical conduit," see Roger Moseley, "Chopin's Aliases," *Nineteenth-Century Music* 42, no. 1 (2018): 3–29, quotation p. 6.
7. Johann Joachim Quantz, *Versuch einer Anweitung die Flôte traversiere zu spielen* (Berlin: Voss, 1752); trans. Edward R. Reily as *On Playing the Flute*, 2nd ed. (London: Faber and Faber, 1985), 131.
8. Scholarship that demonstrates these recurrent practices across the nineteenth and twentieth centuries includes: Katherine Hirt, *When Machines Play Chopin: Musical Spirit and Automation in Nineteenth-Century German Literature* (New York: Walter de Gruyter, 2010); James Q. Davies, *Romantic Anatomies of Performance* (Oakland: University of California Press, 2014); Ritwik Banerji, "Whiteness as Improvisation, Nonwhiteness as Machine," *Jazz & Culture* 4, no. 2 (2021): 56–84; Céline Manning, "Singer-Machines: Describing Italian Singers, 1800–1850," trans. Nicholas Manning, *Opera Quarterly* 28 (2012): 230–258; George Lewis, "Why Do We Want Our Computers to Improvise?," public lecture presented at Monash University (August 13, 2013), https://vimeo.com/78692461; Kelly Hiser, "Electronic Musical Sounds and Material Culture: Early Reception Histories of the Telharmonium, the Theremin, and the Hammond Organ" (PhD thesis, University of Wisconsin–Madison, 2015); Theo Cateforis, *Are We Not New Wave?: Modern Pop*

at the Turn of the 1980s (Ann Arbor: University of Michigan Press, 2011); Sarah Angliss, "Mimics, Menaces, or New Musical Horizons? Musicians' Attitudes Toward the First Commercial Drum Machines and Samplers," in *Material Culture and Electronic Sound*, ed. Frode Weium and Tim Boon (Washington, DC: Smithsonian Institution, 2013), 99–108; Mina Yang, "East Meets West in the Concert Hall: Asians and Classical Music in the Century of Imperialism, Post Colonialism, and Multiculturalism," *Asian Music* 38, no. 1 (2007): 1–30.

9. John Blacking, *How Musical Is Man?* (Seattle: University of Washington Press, 1974), 27.

10. Blacking, *How Musical Is Man?*, 26. Also see critiques of Blacking's use of *human* to distinguish musical and nonmusical sound in Ana María Ochoa Gautier, *Aurality: Listening and Knowledge in Nineteenth-Century Colombia* (Durham, NC: Duke University Press, 2014), 32–75; and Gavin Steingo, "Whale Calling," *Ethnomusicology* 65, no. 2 (2021): 350–373.

11. J. C. R. Licklider, "Man–Computer Symbiosis," *IRE Transactions on Human Factors in Electronics* HFE-1 (March 1960): 4–11.

12. George Lewis, "Why Do We Want Our Computers to Improvise?," in *Oxford Handbook of Algorithmic Music*, ed. Alex McLean and Roger T. Dan (New York: Oxford University Press, 2018), 128.

13. Joel Chadabe, *Electric Sound: The Past and Promise of Electronic Music* (Upper Saddle River, NJ: Prentice Hall, 1997), 291, discussed in Lewis, "Why Do We Want Our Computers to Improvise?," 124.

14. Lucy Suchman, *Human-Machine Reconfigurations* (Cambridge, UK: Cambridge University Press, 2006), 2.

15. Lisa Nakamura, "Prospects for a Materialist Informatics: An Interview with Donna Haraway," *electronic book review*, August 30, 2003.

16. Jacques Barzun, "Introductory Remarks to a Program of Works Produced at the Columbia-Princeton Electronic Music Center," in *Audio Culture: Readings in Modern Music*, ed. Christoph Cox and Daniel Warner (New York: Bloomsbury Academic, 2004), 369.

17. Annette Richards, "Automatic Genius: Mozart and the Mechanical Sublime," *Music & Letters* 80, no. 3 (1999): 366–389, quotations pp. 379, 385.

18. Bonnie Gordon, "*L'Orfeo* at 400: Orfeo's Machines," *Opera Quarterly* 24, no. 3-4 (2009): 200–222; Rebecca Cypess, *Curious and Modern Inventions: Instrumental Music as Discovery in Galileo's Italy* (Chicago: University of Chicago Press, 2016). Jessica Gabriel Peritz identifies a shift from Orpheus's lyre to his voice as the source of his music's powers "in the twilight of the Italian Enlightenment" (around 1770); see her "Orpheus's Civilising Song, or, the Politics of Voice in Late Enlightenment Italy," *Cambridge Opera Journal* 31, no. 2-3 (2020): 129–152, quotation p. 137.

19. Donna Haraway, "A Manifesto for Cyborgs: Science, Technology, and Socialist Feminism in the 1980s," *Socialist Review* 15, no. 2 (1985): 66.

20. N. Katherine Hayles, *How We Became Posthuman: Virtual Bodies in Cybernetics, Literature, and Informatics* (Chicago: University of Chicago Press, 1999).

21. Kodwo Eshun, *More Brilliant Than the Sun: Adventures in Sonic Fiction* (London: Quartet Books, 1998), 06[086].

22. Geer Lovink, "An Interview with Kodwo Eshun by Geer Lovink—Originally from 2000, Republished by Blowup Reader 7 (2013)," V2_, Lab for the Unstable Media, Rotterdam, https://v2.nl/archive/articles/an-interview-with-kodwo-eshun.

23. Alexander G. Weheliye, "'Feenin': Posthuman Voices in Contemporary Black Popular Music," *Social Text* 71 (2002): 21–47; Joseph Auner, "'Sing It for Me': Posthuman Ventriloquism in Recent Popular Music," *Journal of the Royal Musical Association* 128 (2003): 98–122.

24. Roger Grant, "Editorial," *Eighteenth-Century Music* 17, no. 1 (2020): 5.

25. For an overview of historical and scholarly conceptions of machines in relation to music and the human, see Deirdre Loughridge, "'Always Already Technological': New Views of Music and the Human in Musicology and the Cognitive Sciences," *Music Research Annual* 2 (2021): 1–22, ISSN 2563-7290.

26. Recent work developing methods to balance intimate engagement with particular time periods and transhistorical attention to what persists across time includes Roger Moseley, *Keys to Play: Music as a Ludic Medium from Apollo to Nintendo* (Oakland: University of California Press, 2016); Emily I. Dolan and Thomas Patteson, "Ethereal Timbres," in *The Oxford Handbook of Timbre*, ed. Emily I. Dolan and Alexander Rehding (Oxford: Oxford University Press, 2021).

27. Clifford Siskin and William Warner, "This Is Enlightenment: An Invitation in the Form of an Argument," *This Is Enlightenment*, ed. Clifford Siskin and William Warner (Chicago: University of Chicago Press, 2010), 4.

28. Francis Bacon, *The New Organon*, ed. L. Jardin and M. Silverthorn (Cambridge, UK: Cambridge University Press, 2000), 33.

29. Bacon, *The New Organon*, 28.

30. Immanuel Kant, "Was ist Aufklärung?," trans. by Lewis White Beck, *The Politics of Truth*, ed. Sylvère Lotringer and Lysa Hochroth (New York: Semiotext(e), 2007), 29, 37.

31. Minsoo Kang, "The Enlightenment Origins of the Mechanistic Imagery of Humanity," in *Vital Matters: Eighteenth-Century Views of Conception, Life, and Death*, ed. Mary Terrall and Helen Deutsch (Toronto: University of Toronto Press, 2012), 149; Minsoo Kang, *Sublime Dreams of Living Machines: The Automaton in the European Imagination* (Cambridge, MA: Harvard University Press, 2011), 6.

32. Jessica Riskin, *The Restless Clock: A History of the Centuries-Long Argument Over What Makes Life Tick* (Chicago: University of Chicago Press, 2016), 150.

33. Ochoa Gautier, *Aurality*, 61.

34. Dylan Robinson, *Hungry Listening: Resonant Theory for Indigenous Sound Studies* (Minneapolis: University of Minnesota Press, 2020), 98.

35. See Laura Mulvey, *Death 24x a Second: Stillness and the Moving Image* (Chicago: University of Chicago Press, 2006); Amanda Lalonde, "The Music of the Living-Dead," *Music and Letters* 96 (2015): 602–607.

36. Masahiro Mori, "The Uncanny Valley: The Original Essay by Masahiro Mori," trans. Karl F. MacDorman and Norri Kageki, *IEEE Spectrum* (June 12, 2012), https://spectrum.ieee.org/the-uncanny-valley.

37. Mulvey, *Death 24x a Second*, 39.

38. Lynn Festa, *Fiction without Humanity: Person, Animal, Thing in Early Enlightenment Literature and Culture* (Philadelphia: University of Pennsylvania Press, 2019), 1.

39. Louis Chude Sokei, *The Sound of Culture: Diaspora and Black Technopoetics* (Middletown, CT: Wesleyan University Press, 2015), 3, 1–2.

40. Eshun, *More Brilliant Than the Sun*, 06[080].

41. Weheliye, "Feenin," 22; Mara Mills, "Media and Prosthesis: The Vocoder, the Artificial Larynx, and the History of Signal Processing," *Qui Parle* 21, no. 1 (2012): 110.

42. Pascal Bussy, *Kraftwerk: Man Machine and Music*, 3rd ed. (London: SAF Publishing, 2005), 175.

43. *Flame Wars: The Discourse of Cyberculture*, ed. Mark Dery (Durham, NC: Duke University Press, 1994), 213–214. Emphases in original.

44. Weheliye, "Feenin,"40.

45. Wendy Carlos, *Wendy Carlos's Clockwork Orange (Complete Original Score)*, East Side Digital CD (2000), Track 2 "March from a Clockwork Orange." The soundtrack album was originally released in 1972.

46. Christine Lee Gengaro, *Listening to Stanley Kubrick: The Music in His Films* (Toronto: Scarecrow Press, 2013), 140.

47. Judith Peraino, "Synthesizing Difference: The Queer Circuits of Early Synthpop," in *Rethinking Difference in Music Scholarship*, ed. Olivia Bloechl, Melanie Lowe, and Jeffrey Kallberg (Cambridge, UK: Cambridge University Press, 2015), 304.

48. Roy Hollingworth, "The Walter Carlos Sonic Boom," *Melody Maker* (September 23, 1972), https://www.rocksbackpages.com/Library/Article/the-walter-carlos-sonic-boom. On negative reactions to the synthesized voice in Carlos's "Ode to Joy" and her personifications of the vocoder, see Lucie Vágnerová, "Sirens/Cyborgs: Sound Technologies and the Musical Body" (PhD thesis, Columbia University, 2016), 129–135. For further reflection on Carlos's work in relation to notions of *human* and *machine*, see Roshanak Kheshti, *Switched-On Bach* (New York: Bloomsbury Press, 2019).

49. Korg VC-10 ad, *Contemporary Keyboard* (November 1978): 79, https://retrosynthads.blogspot.com/2009/10/korg-vc-10-contemporary-keyboard-1978.html.

50. "When I figured how to make [the vocoder] not just speak, but sing, it earned an assured place in a forthcoming new album. You know, it seemed an exciting idea to share! The first reactions were unanimous: everyone hated it! A playing synth was bad enough, but a 'singing' synth? Too much, turn it off! Thus Timesteps was born, to 'ease into' the first experience most folks would have with a 'singing machine.'" Wendy Carlos interviewed by Kurt B. Reighly for *CMJ New Music Monthly*, "Vocoder Questions" (1999), https://www.wendycarlos.com/vocoders.html.

51. A precedent for Hendryx's first-person "Mother Nature" is Stevie Wonder's "Race Babbling" (1978), in which vocoder-voiced plants caution human listeners, "Can't you see that / Life's connected / You need us to live / But we don't need you." See Francesca T. Royster, "Stevie Wonder's 'Quare' Teachings and Cross-Species Collaboration in *Journey through the Secret Life of Plants* and Other Songs," *Sounding Like a No-No: Queer Sounds and Eccentric Acts in the Post-Soul Era* (Ann Arbor: University of Michigan Press, 2012), 67–87. On Hendryx (with characterization of "Transformation" as robotic on p. 110), see Sonnet Retman, "Between Rock and a Hard Place: Narrating Nona Hendryx's Inscrutable Career," *Women &*

Performance: A Journal of Feminist Theory 16, no. 1 (2006): 107–118; on her prior work with LaBelle, see Gayle Murchison, "Let's Flip It! Quare Emancipations: Black Queer Traditions, Afrofuturisms, Janelle Monáe to Labelle," *Women and Music: A Journal of Gender and Culture* 22 (2018): 79–90.

52. Daphne Oram, *An Individual Note* (London: Galliard, 1972), 123. Tara Rodgers's citation led me to Oram's book, and Rodgers's essay is one of many excellent reflections on archives and historiography: "Tinkering with Cultural Memory: Gender and the Politics of Synthesizer Historiography," *Feminist Media Histories* 1, no. 4 (2015): 5–30. The Daphne Oram Archive has been housed at Goldsmiths, University of London, since 2008.

53. Riskin, *Restless Clock*, 116.

54. Quoted in Jomarie Alano, "The Triumph of the *bouffons*: *La serva padrona* at the Paris Opera, 1752–1754," *The French Review* 79, no. 1 (2005): 127.

55. Wye Jamison Allanbrook, *Comic Mimesis in Late Eighteenth-Century Music*, ed. Mary Ann Smart and Richard Taruskin (Oakland: University of California Press, 2014), 7.

56. Denis Diderot, *Rameau's Nephew: Le Neveu de Rameau*, ed. Marian Hobson, trans. Kate E. Tunstall and Caroline Warman (Cambridge, UK: Open Book Publishers, 2016).

57. Quoted in Roger Grant, "Peculiar Attunements: Comic Opera and Enlightenment Mimesis," *Critical Inquiry* 43 (2017): 562.

58. The finale discussed here, "Per te ho io nel core," was first performed as part of *La serva padrona* in Milan in 1738; it was taken from Pergolesi's *Il Flaminio* (1735) and replaced the original final duet for *La serva padrona*, "Contento tu sarai." According to David R. Stockton, "Per to ho" was the finale used in most major theaters; see his "A New English Performing Edition of Pergolesi's *La Serva Padrona*" (PhD thesis, University of Miami, 1999), 17. The score of *La serva padrona* published in Paris by August Le Duc, giving the opera as "represented in Paris in the fall of 1752," uses the "Per te ho" finale.

59. Carolyn Abbate, *In Search of Opera* (Princeton, NJ: Princeton University Press, 2001); Heather Hadlock, *Mad Loves: Women and Music in Offenbach's Les contes d'Hoffman* (Princeton, NJ: Princeton University Press, 2000).

60. Chude Sokei, *The Sound of Culture*, 47. Further, see Matthew Morrison, "Race, Blacksound, and the (Re)Making of Musicological Discourse," *Journal of the American Musicological Society* 72, no. 3 (2019): 781–823.

61. Lily Kass, "'When then will the veil be lifted?': How Translations Obscure Racism in Productions of *The Magic Flute*," paper presented at American Musicological Society annual meeting, November 2021.

CHAPTER 1

1. "*Nous avons eté de la au fluteur, qui m'a fait bien du plaisir.*" Letter to François Antoine Devaux, a friend from her time at the court of Lorraine, September 8, 1739. See English Showalter, *Françoise de Graffigny: Her Life and Works* (Oxford: Voltaire Foundation, 2004), 55–57; Catherine Liu, *Copying Machines: Taking Notes for the Automaton* (Minneapolis: University of Minnesota Press, 2000), 93.

2. Letter to François Antoine Devaux, September 12, 1739. Showalter, *Françoise de Graffigny*, 57.

3. "Le Fluteur," *Mercure de France* (April 1738), 738; Translated in Kang, *Sublime Dreams of Living Machines*, 103.

4. M. de la Soriniere, "Poeme, ou Essai, sur le progress des Sciences & des Beaux Arts," *Mercure de France* (September 1749), 75.

5. Liu, *Copying Machines*, 72; Reed Benhamou, "From Curiosité to Utilité: The Automaton in Eighteenth-Century France," *Studies in Eighteenth-Century Culture* 17 (1988): 93, 96.

6. Georgia Cowart, *The Triumph of Pleasure: Louis XIV & the Politics of Spectacle* (Chicago: University of Chicago Press, 2008), xviii.

7. Robert Darnton, *The Literary Underground of the Old Regime* (Cambridge, MA: Harvard University Press, 1982), 2.

8. Julien Offray de La Mettrie, *L'Homme machine* (1747), trans. in *Machine Man and Other Writings*, ed. Ann Thomson (Cambridge, UK: Cambridge University Press, 1996).

9. Lorraine Daston and Katherine Park, *Wonders and the Order of Nature* (New York: Zone Books, 1998), 284. Jessica Riskin, *The Restless Clock: A History of the Centuries-Long Argument over What Makes Living Things Tick* (Chicago: University of Chicago Press, 2016), 116.

10. As Riskin notes, Vaucanson's flute player became the "paradigm of an android" when it was used as the defining example in the *Encyclopédie* article "Androïde," published in 1751; see Riskin, *Restless Clock*, 119.

11. Showalter, *Françoise de Graffigny*, quotations pp. xvi and xv. See also Ruth P. Thomas, "Françoise de'Issembourg de Graffigny (11 February 1695–21 December 1758)," *Writers of the French Enlightenment I*, ed. Samia I. Spencer, vol. 313 (Detroit: Gale, 2006), 231–240.

12. Adelheid Voskuhl, *Androids in the Enlightenment* (Chicago: University of Chicago Press, 2013), 170.

13. Lynn Festa, *Sentimental Figures of Empire in Eighteenth-Century Britain and France* (Baltimore, MD: Johns Hopkins University Press, 2006), 43.

14. "*C'est la Représentation en bois, d'un Faune de grandeur naturelle, élevé sur un piédestal proportionné, mais cependant un peu fort, pour contenir toute une manoeuvre aussi compliquée que celle dont on va donner l'explication; tout l'exterieur est peint en couleur de Marbre blanc.*" "Le Fluteur," 738.

15. Jacques Vaucanson, *Le Mécanisme du Fluteur automate, Presenté a Messieurs de L'Académie Royale des Sciences* (Paris: Jacques Guerin, 1738).

16. Paul Vitry, *Antoine Coysevox: Catalogue raisonné de son oeuvre*, vol. 2 (Paris: [G. Keller-Dorian], 1920), 58. Coysevox's statue has also been called *Shepherd Playing the Flute* (Le Berger Jouant de la Flute), a title employed by Riskin. The primary sources around Vaucanson's flute player when it was new, however, consistently refer to the figure as a faun.

17. Cowart, *Triumph of Pleasure*, 30.

18. Jean-Baptiste Fermel'huis, *Eloge funèbre Monsieur Coysevox* (Paris: Collombat, 1721), 20.

19. Antoine Nicolas Dézallier D'Argenville, *Voyage pittoresque des environs de Paris* (Paris: De Bure, 1752).

20. According to the credits printed under the image, it was drawn by H. Gravelot and engraved by Vivares. Born in Paris, Gravelot lived in London from 1732 to 1745. Born near Montpellier, Vivares lived in London from 1727 until his death in 1780. That the artists were in London raises questions about the making of the image I have not been able to answer.

21. Daniel Cottom, "Art in the Age of Mechanical Digestion," *Representations* 66 (1999): 55.

22. Fermel'Huis, *Eloge funèbre Monsieur Coysevox*, 20.

23. Dézallier D'Argenville, *Voyage pittoresque*.

24. Jacques Hotteterre's *Principes de la Flûte Traversière, de la Flûte à Bec, et du Haut-Bois* (Amsterdam: Estienne Roger, 1707), preface.

25. Trans. in James R. Anthony, *French Baroque Music from Beaujoyeulx to Rameau* (Oxford: Oxford University Press, 1973), 394. The literature on French Baroque music cites the French edition, J. C. Nemeitz, *Le Séjour de Paris* (1727), 70–71, but the German edition places Nemeitz's comments a decade earlier: [Joachim Christoph Nemeitz], *Sejour de Paris, oder, Getreue Anleitung* (Frankfurt am Main: Friederich Wilhelm Förster, 1718), 58. Nemeitz was court master to the princely House of Waldeck. He found in Paris "the best concerts, which one can have every day," and mentions attending concerts that were hosted on a regular basis by the Duke D'Aumont, Abbé Grave, Mademoiselle de Maes, Monsieur Clérambault, and Mesdemoiselles Ecuiers; he also mentions Damoiselle la Guerre's concerts having ceased some years ago (57, 69–70).

26. "*Blavet, dont ce Faune a été le disciple.*" Guillaume François Fouques Deshayes Desfontaines, "Lettre CLXXX," in *Observations sur les écrits modernes*, vol. 12 (Paris: Chez Chaubert, 1738), 338; partially in Catherine Cardinal, "Preface," trans. Mary Hyman in *Le Mécanisme du Flûteur Automate* (Paris: Éditions des archives contemporaines, 1985), xvii. The Nightingale of Blavet remains as yet unidentified.

27. Trans. in Dennis Des Chene, *Spirits and Clocks: Machine and Organism in Descartes* (Ithaca, NY: Cornell University Press, 2001), 7.

28. Thomas Tolley, *Painting the Cannon's Roar: Music, the Visual Arts and the Rise of an Attentive Public in the Age of Haydn, c. 1750 to c. 1810* (New York: Routledge, 2001), 156.

29. J. L. Carr, "Pygmalion and the Philosophes: The Animated Statue in Eighteenth-Century France," *Journal of the Warburg and Courtauld Institutes* 23, no. 3-4 (1960): 239–255, quotation at 240.

30. On Amphion, see Wendy Heller, "Dancing Statues and the Myth of Venice: Ancient Sculpture on the Opera Stage," *Art History* 33, no. 2 (2010): 305.

31. Jane Bowers, "A Catalogue of French Works for the Transverse Flute 1692–1761," *Recherches sur la musique française classique* 28 (1978): 89–125.

32. The 1694 edition of the *Dictionnaire de l'Académie française* attests to the range of usages of "Empire" that have relevant meanings here, including power and authority over others, command over one's own passions, and the specific rule of a monarch or emperor; to these, the 1718 edition added reference to a

people. These multiple connotations converge in this scene with the idea of flute and love drawing solitary people from the woods and uniting them under a single, civilizing power. *Dictionnaire de l'Académie française*, 1st ed. (Paris 1694) s.v. "empire"; *Dictionnaire de l'Académie française*, 2nd ed. (1718), s.v. "empire." Thank you to Ellen Lockhart, Scott Sanders, and Chris Parsons for discussing this topic.

33. Georgia Cowart, "Watteau's *Pilgrimage to Cythera* and the Subversive Utopia of the Opera-Ballet," *Art Bulletin* (2001): 465.

34. Michel De La Barre, *Triomphe des Arts* (Paris: Ballard, 1700), 22–23 (act 1).

35. La Barre, *Triomphe des Arts*, 200–201.

36. Letter to François Antoine Devaux dated March 27, 1740. *L'Oracle* was premiered by the Comédiens François Ordinaires du Roi on March 22, 1740, and published that same year. The play also had quite a run outside France. It was translated into English by Susannah Cibber and performed at the Theatre-Royal at Covent Garden in 1763. It was adapted as a German singspiel, *Das Orakel*, with a libretto by Christian Fürchtegott Gellert in 1747; there is evidence of a lost setting by Johann Adam Hiller, and a 1771 setting by Friedrich Gottlob Fleischer survives. See Thomas Baumann, *North German Opera in the Age of Goethe* (Cambridge, UK: Cambridge University Press, 1985), 85–88. David Buch additionally found evidence of a performance of *Das Orakel* in Prague with music by Anton Laube and in Hamburg in 1788; David Buch, *Magic Flutes & Enchanted Forests: The Supernatural in Eighteenth-Century Musical Theater* (Chicago: University of Chicago Press, 2008), 263.

37. [Germain-François Poullain de Saint-Foix], *L'Oracle* (1740) in *Nouveau Theatre François*, vol. 6 (Paris: Prault fils, 1743), 6.

38. De Saint-Foix, *L'Oracle*, 6–7.

39. De Saint-Foix, *L'Oracle*, 8.

40. De Saint-Foix, *L'Oracle*, 9.

41. "*Croyez-vous aussi que votre Clavecin, ou votre Basse de Viole, vous entendent, vous répondent, & sont sensibles aux doux accens de votre voix, lorsqu'ils s'accordent si juste aux tons que vous prenez?*" De Saint-Foix, *L'Oracle*, 9.

42. De Saint-Foix, *L'Oracle*, 9.

43. One might think again here of Descartes, who, describing the animal-machine (*bête-machine*) in *Discours de la méthode* (1637), explained that nature "works within them according to the disposition of their organs, just as one sees a clock, which is made up only of wheels and springs, count the hours." On what this meant for the capacity of animals to feel and have consciousness of their feelings, Descartes remained agnostic. His follower Nicolas Malebranche, however, drew the conclusion that animals were clocklike in their lack of feeling, extrapolating that "they eat without pleasure, cry without pain, grow without knowing it; they desire nothing, fear nothing, know nothing." For Malebranche, feeling required a soul, which God had granted only to humans. See Peter Harrison, "Descartes on Animals," *The Philosophical Quarterly* 42, no. 167 (1992): 219–227.

44. De Saint-Foix, *L'Oracle*, 10.

45. De Saint-Foix, *L'Oracle*, 10.

46. Ellen Lockhart, *Animation, Plasticity and Music in Italy, 1770–1830* (Oakland: University of California Press, 2017), 25.

47. De Saint-Foix, *L'Oracle*, 10–11.

48. On "corporeal sensibility" and "imaginative sensibility," see Alexander Cook, "Feeling Better: Moral Sense and Sensibility in Enlightenment Thought," 85–103, and Henry Martyn Lloyd, "Sensibilité, Embodied Epistemology, and the French Enlightenment," 171–193, in *The Discourse of Sensibility: The Knowing Body in the Enlightenment*, ed. Henry Martyn Lloyd (New York: Springer, 2013).

49. De Saint-Foix, *L'Oracle*, 18–19.

50. Jean-Louis-Ignace de La Serre, *Pirame et Thisbé, tragédie* (Paris: Delormel & Fils, 1759).

51. About two months after seeing *L'Oracle*, Graffigny met the actress who played Lucinde, Jeanne Quinault, and as Showalter writes, the two women developed a "deep and genuine friendship." Showalter, *Françoise de Graffigny*, 60.

52. Festa, *Sentimental Figures*, 45.

53. Vaucanson's automata were in London from December 1742 through July 1743, their run traceable through issues of *The Daily Advertiser*.

54. This sale took place on February 12, 1743; see André Doyon and Lucien Liaigre, *Jacques Vaucanson, mécanicien de genie* (Paris: Presses universitaires de France, 1967), 87–90.

55. See Doyon, *Jacques Vaucanson*, 91. It is worth noting the discrepancy between the text of the poster and the title under which it is catalogued in the Bibliothèque nationale de France, which is: *Jouets Mécaniques de Vaucanson; Un Sauvage, un berger provencal et un canard*.

56. *Dictionnaire de l'Académie français*, 1st ed. (1694), s.v. "sauvage."

57. According to Roger Savage, "To Rameau's contemporaries a *sauvage* was a wild child of nature from any untamed forest. But Frenchmen would be most likely to imagine him as coming from the forests of the Americas, and belonging to one of those tribes with whom they had closest contact in New France, Louisiana and Brazil: tribes such as the Algonquin, the Montagnais, the Huron, the Iroquois, the Natchez and (beyond the equator) the Tupinambá." Roger Savage, "Rameau's American Dancers," *Early Music* 11, no. 4 (1983): 444. See also Michael V. Pisani, *Imagining Native America in Music* (New Haven, CT: Yale University Press, 2008), 37–43.

58. Richard Nash, *Wild Enlightenment: The Borders of Human Identity in the Eighteenth Century* (Charlottesville: University of Virginia Press, 2003), 3.

59. Sophie White, *Wild Frenchmen and Frenchified Indians: Material Culture and Race in Colonial Louisiana* (Philadelphia: University of Pennsylvania Press, 2012), 2. See also Christopher M. Parsons, *A Not-So-New World: Empire and Environment in French Colonial North America* (Philadelphia: University of Pennsylvania Press, 2018).

60. Hélène LeClerc, "*Les Indes galantes* (1735-1952): les sources de l'opéra-ballet; l'exotisme orientalisant; les conditions matérielles du spectacles; fortune des *Indes galantes*," *Revue d'histoire du théâtre Paris* 5, no. 4 (1953): 259–285.

61. Joellen Meglin, "*Sauvages*, Sex Roles, and Semiotics: Representations of Native Americans in the French Ballet, 1736–1837, Part One," *Dance Chronicle* 23, no. 2 (2000): 90.

62. LeClerc finds the costume substitution on inventories for *Les Indes galantes* productions from 1754 at Fontainebleau and 1765 at Versailles. However, the

continued popularity of Rameau's *Les indes galantes* from 1736, together with increased cultural production on the theme exemplified by works like Graffigny's *Lettres d'une Péruvienne* (1747) to be discussed in the next section of this chapter, support the idea that the colonial *sauvage* was a familiar idea to audiences in France at the time of the poster.

63. Jacques Hotteterre, *Principles of the Flute, Recorder & Oboe, by Jacques Hotteterre le Romain*, trans. & ed. David Lasocki (London: Barrie & Rockliff, 1968), 38.

64. Antoine Mahaut, *Nouvelle Méthode pour apprendre en peu de temps a jouer de la flute traversière* (Paris: Lachevardiere, 1759), 4.

65. Hotteterre, *Principles of the Flute, Recorder & Oboe*, 38; Mahaut, *Nouvelle Méthode pour apprendre en peu de temps a jouer de la flute traversière*, 5.

66. Olivia Bloechl, *Native American Song at the Frontiers of Early Modern Music* (Cambridge, UK: Cambridge University Press, 2008), 179.

67. Manuscript completed June 9, 1747. Jo-Ann McEachern and David Smith, "Mme de Graffigny's *Lettres d'une Péruvienne*: Identifying the First Edition," *Eighteenth-Century Fiction* 9, no. 1 (1996): 21–35.

68. On the relation to Montesquieu's *Lettres*, see Martin Calder, *Encounters with the Other: A Journey to the Limits of Language through Works by Rousseau, Defoe, Prévost and Graffigny* (Amsterdam: Rodopi, 2003), 233–235. On interpreting such "Enlightenment ventriloquizations," see Zhuqing (Lester) S. Hu, "Chinese Ears, Delicate or Dull? Toward a Decolonial Comparativism," *Journal of the American Musicological Society* 74, no. 3 (2021): 501–569, quotation 503.

69. Liu, *Copying Machines*, 81.

70. Françoise de Graffigny, *Letters of a Peruvian Woman*, trans. Jonathan Mallinson (Oxford: Oxford University Press, 2009), 3.

71. See Niel Safier, *Measuring the New World: Enlightenment Science and South America* (Chicago: University of Chicago Press, 2008), 200–235; Jonathan Mallinson, "Introduction" to Françoise de Graffigny, *Letters of a Peruvian Woman*, trans. Jonathan Mallinson (Oxford: Oxford University Press, 2009).

72. Lorraine Piroux, "The Encyclopedist and the Peruvian Princess: The Poetics of Illegibility in French Enlightenment Book Culture," *PMLA: Publications of the Modern Language Association of America* 121, no. 1 (2006): 107–123.

73. Graffigny, *Letters of a Peruvian Woman*, 36.

74. Graffigny, *Letters of a Peruvian Woman*, 34.

75. Graffigny, *Letters of a Peruvian Woman*, 44.

76. Graffigny, *Letters of a Peruvian Woman*, 37.

77. Graffigny, *Letters of a Peruvian Woman*, 52.

78. Zilia's music philosophy echoes through Enlightenment texts such as the *Encyclopédie*, where her claims about the universality of music compared to the particularity of language show up in Friedrich Grimm's entry on lyric poetry: "The advantage that the musician's language has over that of the poet is that of a universal language over a specific idiom; the latter only speaks the language of his century and of his country, whereas the former speak the languages of all nations and of all centuries." Friedrich Melchior Grimm, "Lyric poem," *The Encyclopedia of Diderot & d'Alembert Collaborative Translation Project*, trans. Katharina Clausius (Ann Arbor: Michigan Publishing, University of Michigan Library, 2016), accessed

August 19, 2021, http://hdl.handle.net/2027/spo.did2222.0002.817. Originally published as "Poème lyrique," *Encyclopédie ou Dictionnaire raisonné des sciences, des arts et des métiers* 12 (1765): 823–836. On such theories of music and language, see Downing A. Thomas, *Music and the Origins of Language: Theories from the French Enlightenment* (Cambridge, UK: Cambridge University Press, 1995); and Bradley M. Spiers, "Music and the Spectacle of Artificial Life" (PhD diss., University of Chicago, 2020), 76–84.

79. Graffigny, *Letters of a Peruvian Woman*, 52.

80. The balance between language and music was a central issue in Rameau reception; see Charles Dill, *Monstrous Opera: Rameau and the Tragic Tradition* (Princeton, NJ: Princeton University Press, 1998).

81. Trans. in Caroline Wood and Graham Sadler, *French Baroque Opera: A Reader* (London: Taylor & Francis, 2017), 45. From Showalter, *Voltaire et ses amis d'après la correspondance de Mme de Graffigny* I, 1738–1739 SVEC 139 (1975): 198–9.

82. James H. Johnson, *Listening in Paris: A Cultural History* (Berkeley: University of California Press, 1995). Anne Vincent-Buffault, *The History of Tears: Sensibility and Sentimentality in France*, trans. Teresa Bridgeman (New York: St. Martin's Press, 1991).

83. Festa, *Sentimental Figures*, 171.

84. In a second edition of *Lettres d'une Peruvienne*, published in 1752, Graffigny added women to the category of beings whose thinking and feeling is in doubt, having Zilia observe that the French seem to think women don't have souls, that their education is by rote without mind or heart, and that their role is that of "an ornament, there to entertain the curious." Graffigny, *Letters of a Peruvian Woman*, 98–100.

85. Festa, *Sentimental Figures*, 14.

86. I am thinking particularly of the pattern of uncanny disquiet caused by androids established around the second decade of the nineteenth century in the tales of E. T. A. Hoffmann. Most famous among these tales is *The Sandman*: its protagonist Nathaneal falls in love with Olympia, his attraction blossoming when he observes Olympia play the piano and sing. But it turns out that Olympia is an automaton. The revelation drives Nathaneal mad and leaves members of his community with a nagging uncertainty about everyone—are they human or machine?—which directs attunement to and production of bodily signs that specifically differentiate humans from their mechanical simulations (organic vs. clockwork musical time and cough vs. sneeze in *The Sandman*). Such codification of signs that mark one as either human or machine enacts a "human-machine boundary."

87. Carolyn Abbate, "Outside Ravel's Tomb," *Journal of the American Musicological Society* 52, no. 3 (1999): 476.

CHAPTER 2

This chapter is derived in part from an article published in the *Journal of Musicological Research*, 2021 © Taylor & Francis, Deirdre Loughridge, "Metamorphosis and the Taxonomy of Musical Instruments," *Journal of Musicological Research* 40, 3 (2021): 279–288, available online, https://www.tandfonline.com/doi/abs/10.1080/01411896.2021.1949307.

1. Sidney Murray, "Jean-Baptiste Berard's *L'Art du Chant*: Translation and Commentary" (PhD diss., University of Iowa, 1965), 81.
2. Abbate, *In Search of Opera*, 78.
3. Sylvia G. Lenhoff and Howard M. Lenhoff, *Hydra and the Birth of Experimental Biology—1744: Abraham Trembley's Mémoires concerning the Polyps* (Pacific Grove, CA: Boxwood Press, 1986), 6.
4. Lenhoff and Lenhoff, *Hydra and the Birth of Experimental Biology*, 9.
5. Lenhoff and Lenhoff, *Hydra and the Birth of Experimental Biology*, 11.
6. See Marc J. Ratcliffe, *The Quest for the Invisible: Microscopy in the Enlightenment* (New York: Ashgate, 2009), 103–124.
7. Wye Jamison Allanbrook, *Comic Mimesis in Late Eighteenth-Century Music*, ed. Mary Ann Smart and Richard Taruskin (Oakland: University of California Press, 2014), 7.
8. Allanbrook, *Comic Mimesis in Late Eighteenth-Century Music*, 42.
9. Abbate, *In Search of Opera*, 80.
10. On the separation of "human beings" and "nonhumans" into "two entirely distinct ontological zones" as a defining feature of modernity, see Bruno Latour, *We Have Never Been Modern*, trans. Catherine Porter (Cambridge, MA: Harvard University Press, 1993), 10; Jonathan De Souza, "Voice and Instrument at the Origins of Music," *Current Musicology* 97 (2014): 31.
11. Marin Mersenne, *Harmonie universelle*, 4 vols. (Paris: Pierre Ballard, 1636), II, 195; Marin Mersenne, *Harmonie Universelle: The Books on Instruments*, trans. Roger E. Chapman (The Hague: M. Nijhoff, 1957), 254.
12. Gordon J. Kinney, "Trichet's Treatise: A 17th Century Description of the Viols," *Journal of the Viola da Gamba Society of America* 2 (1965): 17.
13. Jamie Savan, "Revoicing a 'Choice Eunuch': The Cornett and Historical Models of Vocality," *Early Music* 46, no. 4 (2017): 561–578.
14. Mersenne, *Harmonie universelle*, II, 195; Mersenne, *Harmonie Universelle*, 254.
15. Le Sieur de Machy, *Pièces de Violle en Musique et en Tablature, differentes les unes des autres, et sur plusieurs Tons* (Paris: Bonneuil, 1685), 11; trans. in John Rutledge, "How Did the Viola da Gamba Sound?," *Early Music* 7, no. 1 (1979): 63.
16. Jean Rousseau, *Traité de la Viole* (Paris: Ballard, 1687), 64; partially trans. in Robert A. Green, "Jean Rousseau and Ornamentation in French Viol Music," *Journal of the Viola da Gamba Society of America* 14 (1977): 19.
17. Gordon J. Kinney, "A 'Tempest in a Glass of Water' or a Conflict of Esthetic Attitudes," *Journal of the Viola da Gamba Society of America* 14 (1977): 44–45.
18. Kinney, "A 'Tempest in a Glass of Water,'" 47–8.
19. "When a man knows his profession well, the chords need not stop him from composing beautiful melodies with all the *agrémens* necessary for playing tenderly" (*les accords ne doivent pas l'embarrasser en composant de beaux chants avec tous les agrémens necessaires pour joüer tendrement*), de Machy, *Pièces de Violle*, 7; trans. in Green, "Jean Rousseau and Ornamentation in French Viol Music," 7.
20. Kinney, "A 'Tempest in a Glass of Water,'" 49. On Rousseau's dispute with de Machy over the viol's proper identity, style, and technique, additionally see Donald Beecher, "Aesthetics of the French Solo Viol Repertory, 1650–1680," *Journal*

of the Viola da Gamba Society of America 24 (1987): 10–21; and Green, "Jean Rousseau and Ornamentation in French Viol Music," 4–21.

21. Rousseau, *Traite de la Viole*, 3, 10–16. A contemporaneous treatise, Danoville's *L'Art de toucher le dessus et basse de violle* (Paris: Ballard, 1687), also maintained that ancient mythology told of an "Orpheus who plays his Lyre with so much sweetness that he causes trees and rocks to move and entices the most ferocious beasts [to follow] after him: These Ancients endeavored to conceal, in the form of this Lyre, the name Viol, or rather, the passing of centuries has changed the name Lyre into that of Viol"; trans. Gordon J. Kinney, "Danoville's Treatise on Viol Playing," *Journal of the Viola da Gamba Society* 12 (1975), 48.

22. According to an engraving made by Abraham Bosse in 1625, the royal gardens at Saint-Germain-en-Laye near Paris, designed by Tommaso and Alessandro Francini, also included a grotto of Orpheus. Saint-Germain-en-Laye provided direct inspiration for Descartes's mechanical analogies for man.

23. Salomon de Caus, *Les Raisons des forces mouvantes*, book II (Frankfurt: Jan Norton, 1615).

24. De Caus does consistently give pride of place to bowed string instruments: Mount Parnassus centers a violin, surrounded by wind and plucked instruments; and there is a scene with Apollo and a "lyre" played with a "bow" (*l'archet*).

25. Rousseau, *Traite de la Viole*, 19. On *tenües*, see Green, "Jean Rousseau and Ornamentation in French Viol Music."

26. Semmens notes that this follows Mersenne's *Harmonie universelle*, III, Proposition IX; see Joseph Sauveur, *Joseph Sauveur's "Treatise on the Theory of Music": A Study, Diplomatic Transcription and Annotated Translation*, trans. Richard Semmens, Studies in Music from the University of Western Ontario 11 (London, Ont.: University of Western Ontario, Faculty of Music, Department of Music History, 1986).

27. Robert Eugene Maxham, "The Contributions of Joseph Sauveur (1653–1716) to Acoustics," vol. II (PhD diss., University of Rochester, Eastman School of Music, 1976), 3.

28. Maxham, "The Contributions of Joseph Sauveur," vol. II.

29. Maxham, "The Contributions of Joseph Sauveur," vol. II, 79. Joseph Sauveur, "Système Général des Intervalles des Sons, et son application à tous les Systèmes et à tous les Instruments de Musique," *Mémoires de l'Académie Royale des Sciences, Année 1701* (Amsterdam: Chez Pierre Mortier, 1736), 405.

30. Maxham, "The Contributions of Joseph Sauveur," vol. II, 124. Sauveur, "Application des sons harmoniques à la composition des Jeux d'Orgues," *Mémoires de l'Académie Royale des Sciences, Année 1702* (Amsterdam: Chez Pierre Mortier, 1736), 23.

31. Maxham, "The Contributions of Joseph Sauveur," vol II, 198; Sauveur, "Rapport des Sons des Cordes d'Instruments de Musique, aux Flêches des Cordes; et nouvelle determination des Sons fixes," *Mémoires de l'Académie Royale des Sciences, Année 1713* (Amsterdam: Chez Pierre Mortier, 1736), 433-469.

32. Alexandra Kieffer, "Bells and the Problem of Realism in Ravel's Early Piano Music," *Journal of Musicology* (2017): 432–472.

33. Hermann von Helmholtz, *On the Sensations of Tone as a Physiological Basis*

for the Theory of Music, trans. Alexander Ellis (London: Longmans, Green and Co., 1885), 70.

34. Jean-Philippe Rameau, *Treatise on Harmony*, trans. Philip Gossett (New York: Dover Publications, 1971). As Thomas Christensen suggests, Rameau must have been unfamiliar with the phenomenon of harmonic overtones perceptible in the sound of a plucked string—perhaps being aware of them only as "strange and apparently inconsistent acoustical by-products"—at the time he published the *Traité*. Otherwise, he surely would have used them in support of his theory, as he thereafter did consistently in his musical writings. Thomas Christensen, *Rameau and Musical Thought in the Enlightenment* (Cambridge, UK: Cambridge University Press, 1993), 133.

35. Rameau, *Treatise on Harmony*, xxxiii.

36. Rameau, *Nouveau système* (Paris, 1726), iii, partially trans. in Christensen, *Rameau and Musical Thought*, 138; also trans. in David Lewin, "Women's Voices and the Fundamental Bass," *Journal of Musicology* 10, no. 4 (1992): 480.

37. Quoted in David E. Cohen, "The 'Gift of Nature': Musical 'Instinct' and Musical Cognition in Rameau," in *Music Theory and Natural Order from the Renaissance to the Early Twentieth Century*, ed. Suzannah Clark and Alexander Rehding (Cambridge, UK: Cambridge University Press, 2001), 77–78.

38. Christensen, *Rameau and Musical Thought*, 13.

39. Rameau, *Observations sur notre instinct pour la musique* (1754), trans. in Jean-Jacques Rousseau, *Essay on the Origin of Languages and Writings Related to Music*, trans. and ed. John T. Scott (Hanover, NH: University Press of New England, 1998), 178.

40. "*L'Harmonie, dans son état primitive & naturel, tel que la donnent les corps sonores, dont notre voix fait partie, doit produire sur nous, qui sommes des corps passivement harmoniques, l'effet le plus naturel, & par consequent le plus commun à tous. De-là vient que celui qui, faute d'une Oreille exercée, est peu sensible aux différentes successions de l'harmonie, l'est au moins par instinct au son d'un corps parfaitement sonore, comme une belle cloche.*" Jean-Jacques Rameau, *Code de musique pratique* (Paris: De L'Imprimerie Royale, 1760), 165.

41. Zhuqing (Lester) S. Hu, "Chinese Ears, Delicate or Dull? Toward a Decolonial Comparativism," *Journal of the American Musicological Society* 74, no. 3 (2021): 501–569.

42. "Mémoires pour server à l'Histoire de la Musique Vocale et Instrumentale," *Mercure de France*, June 1738, 1118.

43. John C. Rule, "The Maurepas Papers: Portrait of a Minister," *French Historical Studies* 4, no. 1 (1965): 104.

44. "*Le Viole tient le juste milieu avec son Archet & ses cordes flexibles, elle donne à la voix sans ôter à la résonnance.*" Hubert Le Blanc, *Defense de la basse de viole contre les enterprises du violon et les pretentions du violoncel* (Amsterdam: Pierre Mortier, 1740), 107; trans. (modified) Barbara Garvey Jackson, "Hubert Le Blanc's *Defense de la viole*," *Journal of the Viola da Gamba Society of America* 10 (1973): 34. On what little is known of Le Blanc's biography, see Jackson's introduction to her translation, "Hubert Le Blanc's *Defense*," 11.

45. Rebekah Ahrendt, "The Diplomatic Viol," in *International Relations, Music and Diplomacy: Sounds and Voices on the International Stage*, ed. Frédéric Ramel and Cécile Prévost-Thomas (Cham, Switzerland: Palgrave, 2018), 102.

46. Don Fader, "The Goûts-réunis in French Vocal Music (1695–1710): Through the Lens of the recueil d'airs sérieux et à boire," *Revue de Musicologie* 96 (2010): 321–363.

47. Emma Spary, *Utopia's Garden: French Natural History from Old Regime to Revolution* (Chicago: University of Chicago Press, 2000), 99.

48. Pierre-Louis Moreau de Maupertuis, *The Earthly Venus*, trans. Simone Brangier Boas (New York: Johnson Reprint Corporation, 1966), 4.

49. William Max Nelson, "Making Men: Enlightenment Ideas of Racial Engineering," *The American Historical Review* 115 (2010): 1371.

50. Mary Terrall, *The Man Who Flattened the Earth: Maupertuis and the Sciences in the Enlightenment* (Chicago: University of Chicago Press, 2002), 199–230.

51. Rousseau, *Traite de la Viole*, 23.

52. Jackson, "Hubert Le Blanc's *Defense*," 19; Le Blanc, *Defense de la basse de viole*, 15.

53. Jackson, "Hubert Le Blanc's *Defense*," 22; Le Blanc, *Defense de la basse de viole*, 25.

54. Antoine Furetière, *Dictionnaire universel*, vol. 1, 1st ed. (Rotterdam: Aroud et Reinier Leers, 1690); Antoine Furetière, *Dictionnaire universel*, vol. 1, 2nd ed. (Rotterdam: Aroud et Reinier Leers, 1701) s.v. deshumaniser. As of the 1740 (3rd) edition, the term still did not appear in the *Dictionnaire de l'Académie française* (Paris, 1740).

55. "*Il ne faut point deshumaniser l'homme en faveur du Heros.*" Henri Basnage de Beauval, "Article X: Extraits de deiverses letter," *Histoire des ouvrages de sçavans*, February 1695, 277.

56. Jackson, "Hubert Le Blanc's *Defense*," 21; Le Blanc, *Defense de la basse de viole*, 23.

57. Jackson, "Hubert Le Blanc's *Defense*," 22; Le Blanc, *Defense de la basse de viole*, 25.

58. Marin Marais, *Pièces de violes... [2e livre]* (Paris: auteur, 1701), 43.

59. On Perrine, *Pièces de luth en musique* (Paris, 1680), see David Buch, "'Style brisé, Style luthé,' and the 'Choses luthées,'" *Musical Quarterly* 71, no. 1 (1985): 52–67.

60. Jackson, "Hubert Le Blanc's *Defense*," 22; Le Blanc, *Defense de la basse de viole*, 26.

61. Michele Mascitti, *Sonate da Camera a violino solo col violone o cembalo*, Op. 2 (Paris: Baussen, 1706).

62. Jackson, "Hubert Le Blanc's *Defense*," 21; Le Blanc, *Defense de la basse de viole*, 21.

63. *Dictionnaire de l'Académie française, vol. 2*, 3rd ed. (Paris: Jean-Baptiste Coignard, 1740), 94.

64. Beecher, "Aesthetics of the French Solo Viol Repertory," 12.

65. See Guillaume Aubert, "'The Blood of France': Race and Purity of Blood in the French Atlantic World," *William and Mary Quarterly* (2004): 439–478.

66. Jennifer Spear, "Colonial Intimacies: Legislating Sex in French Louisiana," *William and Mary Quarterly* LX/1 (2003): 93.

67. Louis Chude Sokei, *The Sound of Culture: Diaspora and Black Technopoetics* (Middletown, CT: Wesleyan University Press, 2015), 133.

68. Vincent Brown, *Tacky's Revolt: The Story of an Atlantic Slave War* (Cambridge, MA: Harvard University Press, 2020), especially 246–249.

69. Chude Sokei, *Sound of Culture*, 78.

70. Barbara Garvey Jackson, "Hubert Le Blanc's *Defense de la viole* (continuation)," *Journal of the Viola da Gamba Society of America* 11 (1974): 21.

71. Examples of this traditional organological system include Sebastian Virdung, *Musica getutscht und ausgezogen* (Basel: [publisher not identified], 1511); and Mersenne, *Harmonie universelle*.

72. Jackson, "Hubert Le Blanc's *Defense de la viole* (continuation)," 22.

73. "... *une Oreille amie du frémissement ou des vibrations du Corps sonore.*" Le Blanc, *Defense de la basse de viole*, 75–76; Jackson, "Hubert Le Blanc's *Defense de la viole* (continuation)," 22.

74. Jackson, "Hubert Le Blanc's *Defense de la viole* (continuation)," 26.

75. Jackson, "Hubert Le Blanc's *Defense de la viole* (continuation)," 24.

76. Jackson, "Hubert Le Blanc's *Defense de la viole* (continuation)," 25. Le Blanc also describes the viol's "modern bow strokes which reproduce and multiply the expressiveness" in contrast to the "old bow strokes... imitating the plucking of the Lute"; Jackson, "Hubert Le Blanc's *Defense de la viole* (continuation)," 41.

77. Jackson, "Hubert Le Blanc's *Defense de la viole* (continuation)," 23.

78. Denis Dodart, *Mémoire sur les causes de la voix de l'homme et de ses différens tons* ([s.n.], 1703), 16.

79. Quoted in Jessica Riskin, "Eighteenth-Century Wetware," *Representations* 83 (2003): 105.

80. Antoine Ferrein, "De la formation de la Voix de l'Homme," *Histoire de l'Académie Royale des Sciences, Année 1741* (Paris: l'Imprimerie Royale, 1744), 410.

81. Antoine Ferrein, "Sur l'Organe immédiat de la Voix et de ses différens tons," *Histoire de l'Académie Royale des Sciences, Année 1741* (Paris: l'Imprimerie Royale, 1744), 51.

82. Ferrein, "De la formation de la Voix," 416; Charles Boulton, trans., *Neurolinguistics Historical and Theoretical Perspectives* (New York: Plenum Press, 1991), 77.

83. Ferrein, "De la formation de la Voix," 412.

84. Ferrein, "De la formation de la Voix," 423.

85. Emily I. Dolan, *The Orchestral Revolution: Haydn and the Technologies of Timbre* (Cambridge, UK: Cambridge University Press, 2013), 58–59.

86. "*Son timbre, d'une haute-contre parfaite, remplissait tellement, que certains sons étaient aussi brillants que s'ils sortaient d'une cloche d'argent.*" *Mémoires de Dufort de Cheverny: La Cour de Louis XV*, ed. Jean-Pierre Guicciardi (Paris: Perrin, 1990), 121.

87. Roseen Giles, "Review Article: A Natural Voice?," *Cambridge Opera Journal* 29 (2018): 250; Martha Feldman, *The Castrato: Reflections on Natures and Kinds* (Oakland: University of California Press, 2015).

88. Daniel Chua, *Absolute Music and the Construction of Meaning* (Cambridge, UK: Cambridge University Press, 1999), 99.

89. J. J. Rousseau, *Dictionnaire de musique*, Tome 1 (Genève, 1781), 143.

90. "*Qui rendent ce travail plus pénible que satisfaisant: car c'est toujours une sotte Musique que celle des Cloches, quand même tous les Sons en seraient exactement justes; ce qui n'arrive jamais.*" Rousseau, *Dictionnaire de musique*, 143. "Which

renders this work more laborious than satisfactory, for the music of bells is silly at the best, tho' all their sounds be exactly true, which never happens." J. J. Rousseau, *A Complete Dictionary of Music*, 2nd ed., trans. William Waring (London: J. Murray, 1779), 55.

91. "*On conçoit que l'extrême gêne à laquelle assujettissent lé concours harmonique des Sons voisins, et le petit nombre des timbres , ne permet guère de mettre du Chant dans un semblable Air.*" Rousseau, *Dictionnaire*, 144. "It is generally imagined that the exceeding difficulties to which the harmonic succession of sounds, and the small number of bells are subject, will not permit anything vocal in an air of this nature." Rousseau, *A Complete Dictionary*, 55. The musical example in the English translation of Rousseau's dictionary contains errors of transcription that introduce seconds into the melody; as seen in the example reproduced here, no melodic seconds are present in the French edition of Rousseau's dictionary.

92. [Claude] Montagnat, *Lettre à M. L'Abbe D. F. Ou Réponse à la Critique que fait M. Burlon, (dans ses Jugemens sur quelques Ouvrages nouveaux) du sentiment de M. F. sur la formation de la voix humaine* (Paris: Davis, Fils, 1745), 22, 21.

93. Montagnat, *Eclaircissemens en forme de letter, a M. Bertin, Medecin, & c, Sur la Découverte que M. Ferrein a faite du mécanisme de la voix de l'homme* (Paris: Davis, Fils, 1746), 16.

94. Anthelme Richerand, *Nouveaux élémens de physiologie* (Paris: Richard, 1801), 465, 467. This text saw numerous editions, in French and in English translation, into the 1830s.

95. Sylvette Milliot, *Historie de la lutherie Parisienne du XVIIIe siècle à 1960*, vol 2: Les luthiers du XVIIIe siècle (Brussels, 2/1997), 152, trans. in Myrna Herzog, "Is the Quinton a Viol? A Puzzle Unravelled," *Early Music* 28, no. 1 (2000): 9.

96. Herzog, "Is the Quinton a Viol?," 22.

97. Haraway, "A Manifesto for Cyborgs," 66.

98. Zakiyyah Iman Jackson, *Becoming Human: Matter and Meaning in an Antiblack World* (New York: New York University Press, 2020), 156.

CHAPTER 3

1. John McCarthy and Claude Shannon, "Preface," *Automata Studies* (Princeton, NJ: Princeton University Press, 1956), v.

2. "A Proposal for the Dartmouth Summer Research Project on Artificial Intelligence" (August 31, 1955), quoted in Rajamaran, "John McCarthy—Father of Artificial Intelligence," *Resonance* 19, no. 3 (2014): 201 [198–207].

3. Norbert Wiener, *Cybernetics: or Control and Communication in the Animal and the Machine* (Cambridge, MA: MIT Press, 1948), 39. Laura Otis similarly narrates: "From the time that investigators first began studying the nervous system, they have described it in terms of contemporary technologies. In the seventeenth and eighteenth centuries, it was a hydraulic system; in the nineteenth, a telegraph net; and in the twentieth, a cybernetic web"; *Networking: Communicating with Bodies and Machines in the Nineteenth Century* (Ann Arbor: University of Michigan Press, 2001), 3.

4. Jacques Barzun, "Introductory Remarks to a Program of Works Produced at the Columbia-Princeton Electronic Music Center," in *Audio Culture: Readings*

in Modern Music, ed. Christoph Cox and Daniel Warner (New York: Bloomsbury Academic, 2004), 369.

5. Remark at the Seventh Macey Conference, March 1950, cited in Slava Gerovitch, *From Newspeak to Cyberspeak: A History of Soviet Cybernetics* (Cambridge, MA: MIT Press, 2002), 90. Following Jessica Riskin, we could think of the slippage from *as if* to *is* idiom in terms of analogy and simulation, which she explains as "two conceptual devices that function quite differently: analogies work by preserving a certain distance between the two things being likened, whereas simulations operate by collapsing that distance." Jessica Riskin, "Eighteenth-Century Wetware," *Representations* 83, no. 1 (2003): 101.

6. N. Katherine Hayles, "Self-Reflexive Metaphors in Maxwell's Demon and Shannon's Choice: Finding the Passages," *Literature and Science: Theory & Practiced*, ed. Stuart Peterfreund (Boston: Northeastern University Press, 1990), 210.

7. Sergi Jorda, "Interactivity and Live Computer Music," *Cambridge Companion to Electronic Music*, ed. Nick Ollins and Julio d'Escrivan (Cambridge, UK: Cambridge University Press, 2017), 92. Thank you to Thomas Patteson for alerting me to this reference.

8. Penelope Gouk, "Clockwork or Musical Instrument? Some English Theories of Mind-Body Interaction before and after Descartes," in *Structures of Feeling in Seventeenth-Century Cultural Expression*, ed. Susan McClary (Toronto: University of Toronto Press, 2013), 35–59. Noting in 1984 that the musical instrument as model for the brain and cognitive functions "despite its iniquitousness, remains unanalysed," Jamie C. Kassler analyzed the model with a focus on "theories of bodily automatisms," particularly "reflex action"; Jamie C. Kassler, "Man—A Musical Instrument: Models of the Brain and Mental Functioning Before the Computer," *History of Science* xxii (1984): 59–92, quotes 59, 60.

9. René Descartes, *The Philosophical Writings of Descartes*, vol. 1, trans. John Cotingham, Robert Stoothoff, and Dugald Murdoch (Cambridge, UK: Cambridge University Press, 1984), 104.

10. Jess Keiser, *Nervous Fictions: Literary Form and the Enlightenment Origins of Neuroscience* (Charlottesville: University of Virginia Press, 2020), 36.

11. See Kassler, "Man—A Musical Instrument." Carmel Raz, "Of Sound Minds and Tuning Forks: Neuroscience's Vibratory Histories," in *The Science-Music Borderlands: Reckoning with the Past and Imagining the Future*, ed. Elizabeth Margulis, Psyche Loui, and Deirdre Loughridge (Cambridge, MA: MIT Press 2023), 115–129.

12. The sensory vibrations potentiate the substance of the brain to vibrate in certain ways, and repeated patterns become more likely. David Hartley, *Observations on Man* (London: S. Richardson, 1749), 56–64.

13. Hartley, *Observations on Man*, 11–12.

14. Hartley, *Observations on Man*, 64.

15. "*Pour se former une idée de l'origine des passions, il est bon de regarder notre individu comme une espèce d'instrument de musique, dont les cordes touchées avec plus ou moins d'accords, donnent des sons plus ou moins harmoniques, & excitent ou le plaisir ou l'ennui. Les nerfs, qui partent, pour la plupart du cerveau, & qui tous au moins s'y rapportent médiatement, se divisent en une infinité de fibres répandues dans toutes les parties du corps, dont ils couvrent même la surface extérieure; ce sont les*

touches & les cordes de l'instrument. Les objets extérieurs qui frappent les extrêmités de ces fibres, y exictent une oscillation plus ou moins forte.... ce sont les mains plus ou moins savants qui touchent l'instrument. L'ame avertie par le reflux des esprits animaux, de tel ou de tel movement, éprouve ou du plaisir ou de la douleur; c'est l'oreille de celui qui écoute l'instrument, & qui se trouve ou agréablement flattée ou creullement déchirée." Armand-Pierre Jacquin, *De la santé, ouvrage utile à tout le monde*, 2nd ed. (Paris: Durand, 1763), 346–347.

16. Denis Diderot, *Diderot's Early Philosophical Works*, trans. Margaret Jourdain (Chicago: Open Court, 1911), 185–186.

17. Riskin, *Restless Clock*, 168.

18. Denis Diderot, *Rameau's Nephew and D'Alembert's Dream*, trans. Leonard Tancock (Harmondsworth, UK: Penguin Classics, 1966), 149. Translations amended.

19. Jean-Jacques Rousseau, *Essay on the Origin of Languages* in *On the Origin of Language*, trans. John H. Moran and Alexander Gode (Chicago: University of Chicago Press, 1986), 64. For more on Rousseau, see Jacqueline Waeber (convenor), "Rousseau in 2013: Afterthoughts on a Tercentenary," *Journal of the American Musicological Society* 66, no. 1 (2013): 251–296.

20. Diderot, *Rameau's Nephew*, 181.

21. Diderot, *Rameau's Nephew*, 149.

22. Diderot, *Rameau's Nephew*, 152.

23. Diderot, *Rameau's Nephew*, 156.

24. Diderot, *Rameau's Nephew*, 156.

25. Diderot, *Rameau's Nephew*, 157.

26. Diderot, *Rameau's Nephew*, 157.

27. Diderot, *Rameau's Nephew*, 157, translation of *clavecin* amended to harpsichord.

28. Erlmann, *Reason and Resonance*, 118, 116.

29. François Couperin, *L'Art de Toucher le Clavecin* (Paris: le sieur Foucault, 1716). Couperin, *Art of Playing the Harpsichord*, trans. Margery Halford (Port Washington, NY: Alfred Publishing, 1974), 33–34.

30. François Couperin, *Pièces de Clavecin, premier livre* (Paris: le sieur Foucault, 1713), 2.

31. Joseph-François-Édouard de Corsembleu de Desmahis, "Woman," *The Encyclopedia of Diderot & d'Alembert Collaborative Translation Project*, trans. Naomi J. Andrews (Ann Arbor: Michigan Publishing, University of Michigan Library, 2004). Accessed January 23, 2023, http://hdl.handle.net/2027/spo.did2222.0000.287. Originally published as "Femme," *Encyclopédie ou Dictionnaire raisonné des sciences, des arts et des métiers* 6 (1756): 472–475.

32. Letter dated July 31, 1762. Denis Diderot, *Oeuvres complètes de Diderot*, vol. XIX, ed. J. Assézat and Maurice Tourneux (Paris: Garnier, 1876), 90–91. Accessed January 23, 2023, https://fr.wikisource.org/wiki/Lettres_%C3%A0_Sophie_Volland/Texte_entier.

33. Diderot, "Man," *The Encyclopedia of Diderot & d'Alembert Collaborative Translation Project*, trans. Nelly S. Hoyt and Thomas Cassirer (Ann Arbor: Michigan Publishing, University of Michigan Library, 2003). Accessed January 23,

2023, http://hdl.handle.net/2027/spo.did2222.0000.160. Originally published as "Homme," *Encyclopédie ou Dictionnaire raisonné des sciences, des arts et des métiers* 8 (1765): 256–257.

34. Desmahis, "Woman."

35. Diderot to Sophie Volland, August 24, 1768, *Correspondence*, vol. 8, ed. Georges Roth (Paris: Éditions de Minuit, 1955), 93. Diderot did not specify harpsichord in this letter, but two weeks later, he mentioned Bayon again, specifically playing harpsichord for him. Her Opus 1, published in 1769, was six sonatas "for harpsichord or pianoforte" (*pour le clavecin ou le piano forte*). See Deborah Hayes, "Marie-Emmanuelle Bayon, Later Madame Louis, and Music in Late Eighteenth-Century France," *College Music Symposium* 30, no. 1 (1990): 14–33.

36. Diderot, *The Indiscreet Jewels*, trans. Sophie Hawkes (New York: Marsilio Publishers, 1993), 124.

37. Diderot, *The Indiscreet Jewels*, 129.

38. Roger Moseley, *Keys to Play* (Oakland: University of California Press, 2016), 91.

39. Thomas McGeary, "Harpsichord Mottoes," *Journal of the American Musical Instrument Society* 7 (1981): 5–35.

40. Sheridan Germann, "Monsieur Doublet and His Confrères: The Harpsichord Decorators of Paris," *Early Music* 8, no. 4 (1980): 435–453.

41. Instrument in The Nydahl Collection, Stockholm, Inventory number IKL059. Accessed January 23, 2023, https://mimo-international.com/MIMO/doc/IFD/OAI_NYC_POST_15056/harpsichord. Also translated (e.g., in McGeary, "Harpsichord Mottoes") as "not unless struck do I sing"; Richard Leppert has noted a recurrent theme of violence in keyboard mottoes, which he relates to the interlinked roles of music and Woman; see "Sexual Identity, Death, and the Family Piano," *19th-Century Music* 16, no. 2 (1992): 105–128.

42. Diderot, *Rameau's Nephew*, 161.

43. Diderot, *Diderot's Early Philosophical Works*, 187.

44. Keiser, *Nervous Fictions*, 20.

45. The only pieces she played that were commercially available at the time were Stockhausen's Studie II and *Gesang der Jünglinge*. The other pieces she obtained through correspondence with studios across Europe and North America.

46. A recording of this lecture is preserved in The Daphne Oram Archive, Special Collections & Archives, Goldsmiths University of London, DO036a. This collection hereafter cited as Oram Archive.

47. Oram Archive, DO036a.

48. Oram Archive, ORAM/4/4/035, 10.

49. Reviews of the event are equally revealing, with headlines like "Groans, Screams and Plops," and "Interesting—but Musical Madness." Oram Archive, ORAM/6/4/002.

50. Oram Archive, DO036b.

51. Oram Archive, DO036b.

52. Oram Archive, ORAM/1/2/006.

53. Oram Archive, ORAM/1/1/002
54. Oram Archive, ORAM/6/5/009.
55. Daphne Oram, "Electronic Music," *The Composer, Journal of the Composers' Guild of Great Britain* 9 (1962): 5–11. In Oram Archive, ORAM/9/04/070.
56. Leopold Stokowski, *Music for All of Us* (New York: Simon and Schuster, 1943).
57. Stokowski, *Music for All of Us*, 12.
58. George Gershwin, "The Composer in the Machine Age (1930)," in *The George Gershwin Reader*, ed. Robert Wyatt and John Andrew Johnson (Oxford: Oxford University Press, 2007), 120. Compare also discourses about mechanical music in 1920s–1930s Germany, discussed in Thomas Patteson, *Instruments for New Music: Sound, Technology and Modernism* (Oakland: University of California Press, 2015).
59. Oram Archive, ORAM/1/2/010, dated March 1961.
60. Oram Archive, ORAM/1/2/039, dated February 4, 1965.
61. Oram Archive, ORAM/1/2/039.
62. Steve Marshall, "Graham Wrench: The Story of Daphne Oram's Optical Synthesizer," *SOS* (February 2009); Tim Richards, "Oramics: Precedents, Technology, and Influence" (Doctoral thesis, Goldsmiths, University of London, 2018), audio 013 (audio recording of interview with Graham Wrench interviewed by Tom Richards and James Bulley, 2012).
63. Oram Archive, ORAM/1/2/070, cited in Richards, "Oramics," 104.
64. Oram Archive, ORAM/4/4/028. For Oramics test audio and discussion of the difficulty of identifying Oramics sound, see Richards, "Oramics," and accompanying audio files 001_Scanner_1965_D0163.mp3 and 002_Scanner_1966.mp3 at https://research.gold.ac.uk/id/eprint/26356/.
65. Oram Archive, DO176.
66. "Tomorrow's World," BBC One, episode presented by Raymond Baxter, produced by Peter Bruce, Clive Parkhurst, and Gordon Thomas, aired March 6, 1968, https://www.bbc.co.uk/archive/peter-zinovieff—early-electronic-music/zmyf7nb.
67. Oram Archive, ORAM/09/04/064.
68. Daphne Oram, "Oramics," *The Magazine of the Institute of Contemporary Arts* 6 (September 1968): 11.
69. Oram, "Oramics," quoting Tristram Cary, "Superserialismus," *Electronic Music Review* 4 (1967).
70. This language echoes the 1960 essay "Man-Computer Symbiosis" by American computer scientist and Macy Conferences participant J. C. R. Licklider.
71. G. L. Mallen, "Art as Process," *Bulletin of the Computer Arts Society* (June 1969): 3.
72. John Landsdown, Goerge Mallen, Alan Mayne, Robert Parslow, Ian Pickering, Jasia Reichardt, Beverley Rowe, Alan Sutcliffe, "Computer Arts Society" Brochure, 1969.
73. Copy of letter to Mr. Braley, December 29, 1970, Daphne Oram Archive, ORAM/6/5/049.
74. Oram, *An Individual Note*, 52.

75. Oram, *An Individual Note*, 6.

76. Robert Kline, "Where Are the Cyborgs in Cybernetics?," *Social Studies of Science* 39, no. 3 (2009): 331–362. Kline argues that analogizing humans and machines was "the main research method of cybernetics," 333–334.

77. Arthur Siegel and Jay Wolf, *Man-Machine Simulation Models: Psychosocial and Performance Interaction* (New York: Wiley-Interscience, 1969), 143.

78. M. Rodger and G. Shapiro, eds., *Prospects for Simulation and Simulators of Dynamic Systems* (New York: Spartan Books, 1967).

79. Slava Gerovitch, *From Newspeak to Cyberspeak: A History of Soviet Cybernetics* (Cambridge, MA: MIT Press, 2002), 94.

80. Canon J. D. Pearce-Higgins and Rev. G Stanely Whitby, eds., *Life, Death and Psychical Research: Studies on behalf of The Churches' Fellowship for Psychical and Spiritual Studies* (London: Rider and Company, 1973), 268–269.

81. Oram, *An Individual Note*, 14.

82. Oram Archive, ORAM/4/7/001, 10.

83. Jeffrey Sconce, *Haunted Media: Electronic Presence from Telegraphy to Television* (Durham, NC: Duke University Press, 2000); Laura Otis, *Networking: Communicating with Bodies and Machines in the Nineteenth Century* (Ann Arbor: University of Michigan Press, 2001). For more on the "scientific-spiritual space" (Andrew Pickering's term) that Oram and cybernetics share with this older tradition, see Deirdre Loughridge, "Daphne Oram, Cyberneticist?," *Resonance: The Journal of Sound and Culture* 2, no. 4 (2021): 503–522.

84. Richard Noakes, "'Instruments to Lay Hold of Spirits': Technologizing the Bodies of Victorian Spiritualism," *Bodies/Machines*, ed. Iwan Rhys Morus (New York: Berg, 2002), 126.

85. Shafica Karagulla, *Breakthrough to Creativity: Your Higher Sense Perception* (Santa Monica: DeVorss, 1967), 26.

86. Karagulla, *Breakthrough to Creativity*, 20.

87. Oram, *An Individual Note*, 35.

88. Oram, *An Individual Note*, 41.

89. Oram, *An Individual Note*, 42.

90. Oram Archive, DO236. A portion of this interview is available as supplemental audio to Richards, "Oramics," audio 011, https://research.gold.ac.uk/id/eprint/26356/. Bird of Parallax was composed for the ballet *Xallaraparallax*, performed by the London Ballet Company.

91. Oram Archive, ORAM/4/4/012.

92. Oram, *An Individual Note*, 100.

93. Oram Archive, DO236.

94. Oram Archive, ORAM/6/5/046. The claim that the violin is a machine also appears in the first edition of *A Dictionary of Music and Musicians*, vol. 4, ed. George Grove (Philadelphia, PA: Theodore Presser, 1895), 268.

95. John Whitney, "Reflections on Art," PAGE, the Bulletin of the Computer Arts Society (March 1972).

96. Rebecca Cypess and Steven Kemper, "The Anthropomorphic Analogy: Humanising Musical Machines in the Early Modern and Contemporary Eras," *Organised Sound* 23, no. 2 (2018): 167–180.

CHAPTER 4

1. Daniel J. Wakin, "For More Pianos, Last Note is Thud in the Dump," *New York Times*, July 30, 2012, A1.

2. Mike J. Dixon, Marie Piskopos, and Tom A. Schweizer, "Musical Instrument Naming Impairments: The Crucial Exception to the Living/Nonliving Dichotomy in Category-Specific Agnosia," *Brain and Cognition* 43 (2000): 159; the musical instruments tested were accordion, harp, guitar, and violin. See also Amy M. Belfi, Anna Kasdan, and Daniel Tranel, "Anomia for Musical Entities," *Aphasiology* 44, no. 3 (2019): 382–404.

3. Jeffrey Sconce, *Haunted Media: Electronic Presence from Telegraphy to Television* (Durham, NC: Duke University Press, 2000), 10.

4. Eliot Bates, "The Social Life of Musical Instruments," *Ethnomusicology* 56, no. 3 (2012): 364.

5. Carolyn Abbate, *In Search of Opera* (Princeton, NJ: Princeton University Press, 2001), 6.

6. Jane Bennett, *Vibrant Matter: A Political Ecology of Things* (Durham, NC: Duke University Press, 2010), 17–18.

7. Bennett, *Vibrant Matter*, xvi.

8. Barbara Johnson, *Persons and Things* (Cambridge, MA: Harvard University Press, 2008), 206–207.

9. Stephen Davies, "What Is the Sound of One Piano Plummeting?," *Themes in the Philosophy of Music* (Oxford: Oxford University Press, 2003), 110.

10. Philip Auslander, "Fluxus Art-Amusement: The Music of the Future?," *Contours of the Theatrical Avant-Garde: Performance and Textuality*, ed. James M. Harding (Ann Arbor: University of Michigan Press, 2000), 126–127.

11. On *The End*, see Ken McLeod, "Living in the Immaterial World: Holograms and Spirituality in Recent Popular Music," *Popular Music and Society* 39, no. 5 (2016): 501–515. For another comparison of (im)mortality in humans and machines, see the 2015 Gob Squad production *My Square Lady* discussed by Gundula Kreuzer, *Curtain, Gong, Steam: Wagnerian Technologies of Nineteenth-Century Opera* (Oakland: University of California Press, 2018), 215–216.

12. "Piano Care," Piano Technicians Guild, accessed March 7, 2017, https://www.ptg.org/ptgmain/piano/care/care-faqs.

13. Frank J. Oteri, "Annea Lockwood Beside the Hudson River," *New Music Box*, January 1, 2004, http://www.newmusicbox.org/articles/annea-lockwood-beside-the-hudson-river/7/.

14. "Piano Drops," Baker House—The Historical Collection, accessed March 7, 2023, http://mit81.com/baker/doku.php?id=baker_spirit:piano_drops.

15. Scott Lipscomb, "Piano Drop—MIT," April 26, 2012, YouTube video, https://www.youtube.com/watch?v=ZECY9M69I5U.

16. Willow Yamauchi, "Years Ago, Canada Produced Beautiful Pianos. Now We Send Them to the Dump," CBC Radio, October 2, 2106, http://www.cbc.ca/radio/thesundayedition/email-madness-ralph-nader-farewell-to-the-upright-piano-gopnik-on-being-a-parent-1.3782876/100-years-ago-canada-produced-beautiful-pianos-now-we-send-them-to-the-dump-1.3782877.

17. Jody Berland, "The Musicking Machine," in *Residual Media*, ed. Charles Acland (Minneapolis: University of Minnesota Press, 2007), 312.

18. Berland, "The Musicking Machine," 312.

19. "Pianos for the Million," *Chambers's Edinburgh Journal* 306 (November 10, 1849): 298.

20. "The Story of a Familiar Friend," *Chambers's Journal of Popular Literature, Science and Arts* 24, no. 95 (October 27, 1855): 260.

21. Edward F. Rimbault, *The Pianoforte, its Origin, Progress and Construction* (London: Robert Cocks and Co., 1860), 161.

22. Sears, Roebuck and Company, *Beckwith Pianos and Player Pianos* (1912), http://antiquepianoshop.com/online-museum/sears-roebuck-company/.

23. Albert Shaw, ed., *The American Review of Reviews* xxxix (January–June 1909).

24. "Steinway: The Instrument of the Immortals," *New York Times*, December 11, 1921, 44. These advertisements also appeared in such papers as the *New York Tribune* and the *Literary Digest*.

25. See Ronald V. Ratcliff, *Steinway* (San Francisco: Chronicle Books, 1989), 46.

26. Davies, "What Is the Sound of One Piano Plummeting?," 117.

27. Lydia Goehr, *The Quest for Voice* (Oxford: Oxford University Press, 2002), 120–121.

28. Georg Wilhelm Friedrich Hegel, *Aesthetics: Lectures on Fine Art*, trans. T. M. Knox (Oxford: Oxford University Press, 1975), 957. This passage from Hegel is also discussed by Amanda Lalonde, "The Music of the Living-Dead," *Music and Letters* 96, no. 4 (2015): 609.

29. Goehr, *Quest for Voice*, 122.

30. Bill Brown, "Thing Theory," *Critical Inquiry* 28, no. 1 (2001): 5.

31. Brown, "Thing Theory," 4.

32. Maurizia Boscagli, *Stuff Theory: Everyday Objects, Radical Materialism* (New York: Bloomsbury, 2014), 228.

33. Lalonde, "The Music of the Living-Dead," 607.

34. Bernard Brauchli, *The Clavichord* (Cambridge, UK: Cambridge University Press, 1998), 174.

35. Thomas McGeary, "Harpsichord Mottoes," *Journal of the American Musical Instrument Society* 7 (1981): 8.

36. E. K. Borthwick, "The Riddle of the Tortoise and the Lyre," *Music and Letters* 51, no. 4 (1970): 373.

37. Rebecca Cypess, *Curious and Modern Inventions: Instrumental Music as Discovery in Galileo's Italy* (Chicago: University of Chicago Press, 2016), quotation 13.

38. Peter Wollny, "Introduction" in *C.P.E. Bach: The Complete Works*, series I, vol. 8.1 (Los Altos: Packard Humanities Institute, 2006), xvi–xvii. On anthropomorphic treatments of the clavichord as a confidant and a being with agency, see Annette Richards, *The Free Fantasia and the Musical Picturesque* (Cambridge, UK: Cambridge University Press, 2001), 155–164.

39. According to Samuel von Pufendorf's *De iure naturae et gentium libri octo* (1672), which remained influential into the mid-eighteenth century, a necessary condition of marriage was that the bride "must have come, as it were, into the

hands of the man in such a way that he may treat her as a wife." See Charles Donaheu, "The Legal Background: European Marriage Law from the Sixteenth to the Nineteenth Century," in *Marriage in Europe, 1400–1800*, ed. Silvana Seidel Menchi (Toronto: University of Toronto Press, 2016), 49.

40. Nannette Streicher, *Kurze Bemerkungen uber das Spielen, Stimmen und Erhalten der Fortpiano* (Vienna: Abertischen Schriften, 1801), trans. as *Brief Remarks on the Playing, Tuning, and Maintenance of Fortepianos* in Preethi De Silva, *The Fortepiano Writings of Streicher, Dieudonné, and the Schiedmayers* (Lewiston, NY: Edwin Mellen Press, 2008), 60–61.

41. De Silva, *The Fortepiano Writings of Streicher, Dieudonné, and the Schiedmayers*, 63.

42. See Dana Gooley, *The Virtuoso Liszt* (Cambridge, UK: Cambridge University Press, 2004), 106–113.

43. "The Story of a Familiar Friend," 260.

44. Algernon Rose, ed., *A439: Being the Autobiography of a Piano* (London: Sands & Co., 1900), 23.

45. Rose, *A439*, 24.

46. Rose, *A439*, 25.

47. Rose, *A439*, 19.

48. Siegfried Hansing, *The Pianoforte and Its Acoustic Properties*, 2nd ed., trans. Emmy Hansing-Perzina (New York: Schwerin, 1904), 96–103.

49. Descartes, *The Philosophical Writings of Descartes*, vol. 3, The Correspondence, 143, quoted and discussed in Keiser, *Nervous Fictions*, 23.

50. Albert Shaw, ed., "Where Art Is Greater Than Science," *The American Review of Reviews* 39 (1909), 67.

51. Franz Mohr, chief concert technician for Steinway from 1968 to 1992, exemplifies crediting the Steinway soundboard with the instrument's wide range of "tonal possibilities." Franz Mohr, *My Life with the Great Pianists* (Grand Rapids, MI: Baker Books, 1996), 126; Mohr also expresses this idea in the documentary *Note by Note: the Making of Steinway L1037* (2007).

52. Shaw, ed., "A Secret of Home," *The American Review of Reviews* 39 (1909), 47.

53. [Alexander Nikitich] Bukhovstev, *Guide to the Proper Use of the Pianoforte Pedals with Examples out of the Historical Concerts of Anton Rubinstein*, trans. John A. Preston (New York: Bosworth & Co., 1897), title page; reproduced in *The Art of Piano Pedaling: Two Classic Guides* (Mineola: Dover Publications, 2003), 1.

54. Sean Murray, "Pianos, Ivory, and Empire," *American Music Review* XXXVIII, no. 2 (Spring 2009): 4. On the process of making ivory key veneers, see David H. Shayt, "Elephant under Glass: The Piano Key Bleach House of Deep River, Connecticut," *IA The Journal of the Society for Industrial Archeology* 19, no. 1 (1993): 37–59. On the broader context of the ivory trade, see Alexandra Celia Kelly, *Consuming Ivory: Mercantile Legacies of East Africa and New England* (Seattle: University of Washington Press, 2021).

55. Wendy Hui Kyong Chun, "Introduction: Race and/as Technology; or, How to Do Things with Race," *Camera Obscura* 24, no. 1 (70) (2009): 7–35. Beth Coleman,

"Race as Technology," *Camera Obscura* 24, no. 1 (70) (2009): 177–207, quotation at 184.

56. R. Allen Lott, *From Paris to Peoria: How European Piano Virtuosos Brought Classical Music to the American Heartland* (Oxford: Oxford University Press, 2003), 92–95. The source in which "injury" appears is Maude Haywood, "Decorations for the Piano," *The Ladies' Home Journal* 10 (January 1983): 9. As this and other women-oriented publications of the era suggest, "care of the piano" was a responsibility that fell to housewives; Mary Bancroft Smith, *Practical Helps for Housewives* (Rutherford, NJ: Bergen County Herald Publishing Company, 1896), 40.

57. "The Challenges of Making a Digital Piano," BBC News, April 17, 2014, http://www.bbc.com/news/av/technology-27062526/the-challenges-of-making-a-digital-piano-sound-real.

58. ILIO, "Ivory II—Grand Pianos: What Makes It Different," November 19, 2015, YouTube video, https://www.youtube.com/watch?v=ieQ7osfDLAw.

59. "The Challenges of Making."

60. Yamaha Corporation, "Yamaha AvantGrand N3X: Interview with Francesco Tristanto," October 3, 2016, YouTube video, https://www.youtube.com/watch?v=QR05aRXlz9s.

61. "No Strings Attached," *The Economist*, February 27, 2009, http://www.economist.com/node/13208736.

62. N. Katherine Hayles, *How We Became Posthuman* (Chicago: University of Chicago Press, 1999), 1, xi.

63. Examples of such films include *The Love of Life* (1969); *Mozart in Japan* (1987); and *Thelonius Monk: Straight, No Chaser* (1988).

64. Examples include "Matt Sorum Drums Up Musicians' Support for US Ivory Ban," *PR Newswire*, June 3, 2014; Sarah Bryan Miller and Chuck Raasch, "Ivory Ban Hurts Musicians, Collectors," *TCA Regional News*, June 22, 2014; and "Feds and Musicians Face Off on Ivory Law: Mondomusica New York Brings APHIS and US Fish and Wildlife Agents to a Seminar on New Ivory Transport Enforcement," *Music Trades* 162, no. 4 (May 2014), 36.

65. "Interior Announces Ban on Commercial Trade of Ivory as Part of Overall Effort to Combat Poaching, Wildlife Trafficking," US Department of the Interior Press Releases, February 11, 2014, https://www.doi.gov/news/pressreleases/interior-announces-ban-on-commercial-trade-of-ivory-as-part-of-overall-effort-to-combat-poaching-wildlife-trafficking.

66. Edmond Johnson, "Policies Intended to Protect Elephants May Inadvertently Endanger Musical Instruments," *Newsletter of the American Musical Instrument Society* 43, no. 2 (2014): 1, 8–10.

67. "Frequent Cross-Border Non-commercial Movements of Musical Instruments," Convention on International Trade in Endangered Species of Wild Fauna and Flora, Conf. 16.8, accessed January 24, 2023, https://cites.org/sites/default/files/document/E-Res-16-08-R17_0.pdf.

68. Ben Niles, dir., *Note by Note: The Making of Steinway L1037* (New York: Docurama, 2007).

69. Jane Bennett, *Vibrant Matter: A Political Ecology of Things* (Durham, NC: Duke University Press, 2010), 107.

CHAPTER 5

1. Chris Richards, "There Is No Song of the Summer," *Washington Post*, July 28, 2016, Academic OneFile, accessed July 11, 2018, http://link.galegroup.com/apps/doc/A459391029/AONE?u=mlin_b_northest&sid=AONE&xid=45c9825b.

2. Robby Seabrook III, "The Evolution of Chopped and Repitched Samples in Music," Genius, September 1, 2017, https://genius.com/a/the-evolution-of-chopped-repitched-samples-in-pop-music. The pitch-shifting in "chopped and pitched" vocals may be audible (as when extreme pitch-shifting affects the timbre) or inaudible (subtle pitch "correction"). Similarly, the *chop* in "chopped and pitched" vocals refers to the production technique known as "sample slicing," which involves separating a relatively brief section out from a longer recording, which may or may not result in sounding "chopped" in the sense of featuring abrupt starts and stops. The vocal chop is shared with the "chopped and screwed" technique named for DJ Screw, who developed the practice in Houston hip-hop in the 1990s; in this music, pitch-shifting is achieved by slowing a recording down and lowering the pitch along with altering tempo and timbre. On the "chopped and screwed" sound and technique, see Lance Scott Walker, *DJ Screw: A Life in Slow Revolution* (Austin: University of Texas Press, 2022); and Langston Collin Wilkins, "Ep. 7: Screwed and Chopped," *Phantom Power* podcast, cohosted by Mack Hagood, June 6, 2018, https://phantompod.org/ep-7-screwed-chopped-langston-collin-wilkins/.

3. Richards, "There Is No Song of the Summer."

4. See Allison McCracken, *Real Men Don't Sing: Crooning in American Culture* (Durham, NC: Duke University Press, 2015).

5. Thomas Porcello, "The Ethics of Digital Audio-Sampling: Engineers' Discourse," *Popular Music* 10 (1991): 77.

6. "Vocal Chopping & Pitching," *Attack Magazine*, October 11, 2012, https://www.attackmagazine.com/technique/tutorials/vocal-chopping-and-pitching/.

7. Susana Loza, "Sampling (Hetero)sexuality: Diva-ness and Discipline in Electronic Dance Music," *Popular Music* 20 (2001): 349.

8. Joseph Auner, "Losing Your Voice: Sampled Speech and Song from the Uncanny to the Unremarkable," in *Throughout: Art and Culture Emerging with Ubiquitous Computing*, ed. Ulrik Ekman (Cambridge, MA: MIT Press, 2013), 142.

9. Jonah Weiner, "Daft Punk Reveal Secrets of New Album," *Rolling Stone*, April 13, 2013, https://www.rollingstone.com/music/music-news/daft-punk-reveal-secrets-of-new-album-exclusive-191153/.

10. "Video: Justin Bieber, Diplo and Skrillex Make a Hit," *New York Times*, August 25, 2015, https://www.nytimes.com/interactive/2015/08/25/arts/music/justin-bieber-diplo-skrillex-make-a-hit-song.html.

11. "DJ Snake Humbled by Instant Success of Star-Studded 'Encore,'" KYXY Radio, August 10, 2016, http://kyxy.radio.com/2016/08/10/dj-snake-humbled-by-instant-success-of-star-studded-encore/.

12. Suzanne Cusick, "On Musical Performances of Gender and Sex," in *Audible Traces: Gender, Identity, and Music*, ed. Elaine Barkin and Lydia Hanessley (Los Angeles: Carciofoli, 1999), 25–48; Nina Eidsheim, "'Sonic Blackness' in American

Opera," *American Quarterly* 63 (2011): 641–671; James Q. Davies, *Romantic Anatomies of Performance* (Berkeley: University of California Press, 2014); Ana María Ochoa Gautier, *Aurality: Listening and Knowledge in Nineteenth-Century Colombia* (Durham, NC: Duke University Press, 2014); Alisha Lola Jones, *Flaming? The Peculiar Theopolitics of Fire and Desire in Black Male Gospel Performance* (Oxford: Oxford University Press, 2020); Jennifer Lynn Stoever, *The Sonic Color Line: Race and the Cultural Politics of Listening* (New York: NYU Press, 2016); Kira Thurman, *Singing Like Germans: Black Musicians in the Land of Bach, Beethoven, and Brahms* (Ithaca, NY: Cornell University Press, 2021).

13. Sylvia Wynter, "Unsettling the Coloniality of Being/Power/Truth/Freedom: Towards the Human, After Man, Its Overrepresentation—An Argument," *CR: The New Centennial Review* 3, no. 3 (2003): 330. On "genres of being human," see also Sylvia Wynter and Katherine McKittrick, "Unparalleled Catastrophe for Our Species? Or, to Give Humanness a Different Future: Conversations," in *Sylvia Wynter: On Being Human as Praxis*, ed. Katherine McKittrick (Durham, NC: Duke University Press, 2015), 9–89. On Sylvia Wynter's critique of "man" and the posthuman in music, see also Weheliye, "Feenin" and Justin Adams Burton, *Posthuman Rap* (Oxford University Press, 2017).

14. David Yearsley, *Bach and the Meanings of Counterpoint* (Cambridge, UK: Cambridge University Press, 2002), 179.

15. Adelheid Voskuhl, *Androids in the Enlightenment* (Chicago: University of Chicago Press, 2013), 200.

16. See Jeffrey Kallberg, "Mechanical Chopin," *Common Knowledge* (2011): 269–282.

17. "*Je me fais l'effet d'une mécahnique à chanter—à 8 h.res du soir on pousse mon ressort et je chante*," Pauline Viardot, letter to Franz Liszt, Newcastle-upon-Tyne, February 17, 1859, published in La Mara, ed., *Briefe hervorragender Zeitgenossen an Franz Liszt*, vol II (Leipzig: Breikopf & Hartel, 1985), 207. Translated by Hilary Poriss and Maurice Bombron. Thanks to Hilary Poriss for bringing this letter to my attention.

18. Céline Manning, "Singer-Machines: Describing Italian Singers, 1800–1850," trans. Nicholas Manning, *Opera Quarterly* 28 (2012): 230.

19. Nicolò Eustachio Cattaneo, *Gazzetta musicale di Milano*, June 5, 1842, cited by Manning, "Singer-Machines," 237.

20. *Memoirs of Hector Berlioz from 1803 to 1865*, trans. Rachel Holmes, Eleanor Holmes, and Ernest Newman (New York: Dover, 1960), 372.

21. See Katharine Ellis, "Female Pianists and Their Male Critics in Nineteenth-Century Paris," *Journal of the American Musicological Society* 50 (1997): 353–385; Heather Hadlock, *Mad Loves: Women and Music in Offenbach's "Les Contes d'Hoffmann"* (Princeton, NJ: Princeton University Press, 2000), esp. 70–71; Susan C. Cook and Judy S. Tsou, eds., *Cecilia Reclaimed: Feminist Perspectives on Gender and Music* (Chicago: University of Illinois Press, 1994).

22. Kurt Beals, "'Do the New Poets Think? It's Possible': Computer Poetry and Cyborg Subjectivity," *Configurations* 26 (2018): 151.

23. Claudia Springer, "Pleasure of the Interface," *Screen* 32 (1991): 306.

24. Haraway, "A Manifesto for Cyborgs," 66.

25. See Devon Powers, *Writing the Record: The Village Voice and the Birth of Rock Criticism* (Amherst: University of Massachusetts Press, 2013).

26. On Joshua Clover/Jane Dark, see Brooke Kroeger, *Passing: When People Can't Be Who They Are* (New York: PublicAffairs, 2003), 167–208, at 197 and 169.

27. Jane Dark, "Humans, Posthumans, and a Detergent Ad," *Village Voice*, November 16, 1999.

28. Tom Breihan, "The Battle for the Heart of R&B," *Village Voice*, June 7, 2007.

29. On Rihanna's reception as a "robo-diva," see Robin James, "Robo-Diva R&B: Aesthetics, Politics, and Black Female Robots in Contemporary Popular Music," *Journal of Popular Music Studies* 20 (2008): 402–423. On Whitney Houston's reception, Charles Aaron, "Nippy's Got the Range: How Whitney Houston Snatched Back the Ballad and Set It Free," paper presented at Pop Conference 2018, Museum of Popular Culture, Seattle, Washington, April 26, 2018. On recognizing craft in the context of soul, see Emily J. Lordi, *The Meaning of Soul: Black Music and Resilience Since the 1960s* (Durham, NC: Duke University Press, 2020).

30. Sasha Frere-Jones, "The Gerbil's Revenge," *New Yorker*, June 9, 2008, 128.

31. Ben Westhoff, *Dirty South: Outkast, Lil Wayne, Soulja Boy, and the Southern Rappers Who Reinvented Hip-Hop* (Chicago: Chicago Review Press, 2011), 214.

32. These viral videos were documented, for example, by ViralViralVideos.com, accessed January 24, 2023, https://www.viralviralvideos.com/2011/04/06/kesha-singing-without-auto-tune/; and *Time*, July 9, 2014, http://time.com/2969757/britney-spears-alien-autotune/. For an analysis of how gender and race impacted the reception of Auto-Tuned voices, see Catherine Provenzano, "Making Voices: The Gendering of Pitch Correction and the Auto-Tune Effect in Contemporary Pop Music," *Journal of Popular Music Studies* 31, no. 2 (2019): 63–84.

33. "T-Pain v. His Vocoder," Funny or Die, September 30, 2008, http://www.funnyordie.com/videos/86a76df842/tpain-v-his-vocoder-from-tpain, reposted as "T-Pain Vs. Vocoder," Creative Destruction, YouTube video, https://www.youtube.com/watch?v=234v_apCQO4.

34. Marc Weidenbaum, "Recording Studios May Die, but the False Mythology Around Them May Not," *The Atlantic*, December 5, 2012, https://www.theatlantic.com/entertainment/archive/2012/12/recording-studios-may-die-but-the-false-mythology-around-them-may-not/265919/.

35. Lessley Anderson, "Seduced by 'Perfect Pitch': How Auto-Tune Conquered Pop Music," *The Verge*, February 27, 2013, https://www.theverge.com/2013/2/27/3964406/seduced-by-perfect-pitch-how-auto-tune-conquered-pop-music.

36. Frere-Jones, "The Gerbil's Revenge," 128.

37. Melissa Goldstein, "Death Cab Wage War Against Auto-Tune," *Spin Magazine*, February 10, 2009, https://www.spin.com/2009/02/death-cab-wage-war-against-auto-tune/.

38. Jessica Riskin, "The Defecating Duck, or, the Ambiguous Origins of Artificial Life," *Critical Inquiry* 29 (2003): 623.

39. Eshun, *More Brilliant Than the Sun*; James, "Robo-Diva"; Jack Hamilton, *Just Around Midnight* (Cambridge, MA: Harvard University Press, 2016).

40. Sankar Muthu, *Enlightenment Against Empire* (Princeton, NJ: Princeton University Press, 2003), 67–68.

41. Kodwo Eshun credits exchanges between *Village Voice* writer Greg Tate and other *Wire* associated writers (Mark Sinker, David Toop) with setting the stage for his book *More Brilliant Than the Sun*, A[175]. He gives Greg Tate priority in developing the concept of Afrofuturism, revising the priority often assigned Mark Dery for coining the term in his introduction to a set of interviews: "Black to the Future: Interviews with Samuel R. Delany, Greg Tate, and Tricia Rose," *South Atlantic Quarterly* 94 (1993), repr. in *Flame Wars: The Discourse of Cyberculture*, ed. Mark Dery (Durham, NC: Duke University Press, 1994).

42. Eshun, *More Brilliant Than the Sun*, A[192]–A[193].

43. Eshun's extensive bibliography includes five Marshall McLuhan titles.

44. Eshun, *More Brilliant Than the Sun*, 00[–002].

45. Jace Clayton, "Pitch Perfect," *Frieze* 123 (May 2009), https://frieze.com/article/pitch-perfect; Clayton quotes Jody Rosen, "Sexbot: Why T-Pain Is the Perfect Web 2.0 Pop Star," *Slate*, July 26, 2007, http://www.slate.com/articles/arts/music_box/2007/07/sexbot.html. Rosen is in fact quite consistent about figuring T-Pain as a cyborg rather than pure robot or computer in this article, and his overall assessment of T-Pain is positive. However, unlike Clayton, Rosen figures T-Pain's voice as pure machine ("Man With the Robot Voice"), and because Rosen did not argue for the cyborg as a positive figure, it is easy to read his various other characterizations of T-Pain—"sex machine," "robo-shtick," C-3PO's voice with dreadlocks—as critical slights.

46. For Ann Powers's earlier work on Auto-Tune, see Powers, "Kanye West Examines Real vs. Fake, Puppet vs. Human on '808s and Heartbreak,'" *LA Times Blog*, November 21, 2008, http://latimesblogs.latimes.com/music_blog/2008/11/kanye-west-exam.html; Powers, "Is the Internet Making Us Love Insular Music?," *Slate*, December 23, 2008, http://www.slate.com/articles/arts/the_music_club/features/2008/the_music_club/is_the_internet_making_us_love_insular_music.html. The latter is a response to Auto-Tune complaints and worries from Jody Rosen and Robert Christgau.

47. Ann Powers, *Good Booty: Love and Sex, Black and White, Body and Soul in American Music* (New York: Dey St., 2017), 302.

48. Powers, *Good Booty*, 303.

49. Powers, *Good Booty*, 332.

50. Simon Reynolds, "How Auto-Tune Revolutionized the Sound of Popular Music," *Pitchfork*, September 17, 2018, https://pitchfork.com/features/article/how-auto-tune-revolutionized-the-sound-of-popular-music/. Reynolds also uses the term *posthuman* but pairs it with *perfection* and *precision* to describe the intended use of Auto-Tune for pitch correction, making it read as a figure of the inhuman machine in contrast to the cyborg. Reynolds's "*machines* and *soulfulness*" is an update to Andrew Goodwin's "*machines* and *funkiness*" in "Sample and Hold: Pop Music in the Digital Age of Reproduction," *Critical Quarterly* 30, no. 3 (1988): 39.

51. Elias Leight, "Skin Companion EP II," *Pitchfork*, February 28, 2017, https://pitchfork.com/reviews/albums/22965-flume-skin-companion-ep-ii/.

52. Chris Richards, "Holly Herndon's Right-Now Sound," *Washington Post*, May 8, 2015.

53. Mark O'Connell, *To Be a Machine: Adventures Among Cyborgs, Utopians,*

Hackers, and the Futurists Solving the Modest Problem of Death (New York: Doubleday, 2017); quoted by Chris Richards, "How the Death of EDM Brought Pop Music One Step Closer to Eternal Life," *Washington Post*, August 3, 2017. By contrast, David Trippet emphasizes transhumanists' interests in genetic modification and implants and designates the fantasized achievement of disembodied mind *the posthuman*, as distinct from *posthumanism*, which he sums up as a wider discursive/philosophical response to humanist anthropocentrism, in "Music and the Transhuman Ear: Ultrasonics, Material Bodies, and the Limits of Sensation," *The Musical Quarterly* 100 (2018): 199–261.

54. Richards, "How the Death of EDM Brought Pop Music One Step Closer."

55. On Wendy Carlos, see Judith Peraino, "Synthesizing Difference: The Queer Circuits of Early Synthpop," in *Rethinking Difference in Music Scholarship*, ed. Oliva Bloechl, Melanie Lowe, and Jeffrey Kallberg (Cambridge, UK: Cambridge University Press, 2015), 287–313; on Laurie Anderson, see Susan McClary, *Feminine Endings* (Minneapolis: University of Minnesota Press, 1991), 132–147. Nona Hendryx made use of a vocoder effect on her 1983 songs "Transformations" and "Living on the Border"; the former is discussed in the introduction to this book.

56. On Grace Jones, see Francesa T. Royster, "Feeling Like a Woman, Looking Like a Man, Sounding Like a No-No," in *Sounding Like a No-No* (Ann Arbor: University of Michigan Press, 2013), 142–165.

57. Kay Dickinson, "'Believe'? Vocoders, Digitalised Female Identity and Camp," *Popular Music* 20 (2001): 333–334.

58. Sue Sillitoe, "Recording Cher's 'Believe': Mark Taylor & Brian Rawling," *Sound on Sound*, February 1999, https://www.soundonsound.com/techniques/recording-cher-believe.

59. Dickinson, "'Believe'?," 343; Nick Prior contemplates this issue in "Software Sequencers and Cyborg Singers," *New Formations* 66 (2009): 93–94.

60. Stacy L. Smith, Marc Choueiti, and Katherine Pieper, *Inclusion in the Recording Studio?*, Annenberg Inclusion Initiative (Los Angeles: USC Annenberg School for Communication and Journalism, 2018).

61. "Feeling This: A Conversation with Grimes," NPR, April 27, 2016, https://www.npr.org/2016/04/27/455769676/feeling-this-a-conversation-with-grimes.

62. Emma Mayhew, "Positioning the Producer: Gender Divisions in Creative Labour and Value," in *Music, Space, and Place*, ed. Sheila Whiteley, Andy Bennett, and Stan Hawkins (Aldershot, UK: Ashgate, 2004), 157.

63. Andy Hildebrand, "A Few Words from Dr. Andy," in Antares, *Auto-tune™: Intonation Correcting Plug-In: User's Manual* (Los Gatos, CA: Antares Audio Technologies, 2000), 7.

64. This absence suggests a piano-based understanding of music, the piano being an instrument that similarly outsources intonation (to a piano tuner), leaving the performer to make expressive decisions outside the domain of pitch. Unsurprisingly, the default interface for Antares Auto-Tune software represents pitches on a keyboard. For more on constructions of emotional versus technical labor in and around Auto-Tune, see Provenzano, "Making Voices," and Marshall Owen, "Auto-Tune In Situ: Articulations of Voice, Affect, and Artifact in the Recording Studio" (PhD diss., Cornell University, 2017).

65. "This Is What You Came For—Calvin Harris invited Radio 1 to LA for a first play of his new single with Rihanna," BBC Radio 1, New Music Friday, April 29, 2016, http://www.bbc.co.uk/programmes/articles/55y7qwFPKZcsYFg93XHzcLz/this-is-what-you-came-for-calvin-harris-invited-radio-1-to-la-for-a-first-play-of-his-new-single-with-rihanna.

66. "Interview: The Rise of Felix Snow," *Kick Kick Snare*, May 16, 2016, https://kickkicksnare.squarespace.com/new-releases/2016/05/16/interview-the-rise-of-felix-snow.

67. "Kiiara Interview—Joey Franchize—WiLD 941," WiLD 94.1, July 11, 2016, YouTube video, https://www.youtube.com/watch?v=MlViZ6jqO9E.

68. Alim Kheraj, "We Got MNEK and Zara Larsson to Interview Each Other and Here Are the Results," *Pop Buzz*, October 29, 2015, http://www.popbuzz.com/music/features/mnek-zara-larsson-interview/#0oGQzQHRm8VESM9H.99.

69. Emily McDermott, "Refracting Sound," *Interview Magazine*, May 19, 2015, https://www.interviewmagazine.com/music/holly-herndon-platform.

70. Stacey Anderson, "*Skin*," *Pitchfork*, June 2, 2016, https://pitchfork.com/reviews/albums/21921-skin/. Special thanks to Grant Foskett for this example and for illuminating discussions of a wider context for this voice-synthesizer phenomenon in electronic dance music.

71. "#TBT 2013: Calvin Harris Reveals How He Linked Up with Rihanna for 'We Found Love,'" *Fuse*, January 12, 2017, https://www.fuse.tv/videos/2017/01/tbt-calvin-harris-rihanna-we-found-love.

72. "What Is Happening at the Beginning of Kiiara's 'Gold' and What Was Her College Major," 95.5. PLJ, August 6, 2016, YouTube video, https://www.youtube.com/watch?v=OJvsbpbQt0g.

73. See Tara Rodgers, *Pink Noises: Women on Electronic Music and Sound* (Durham, NC: Duke University Press, 2010); Lucy Green, *Music, Gender, Education* (Cambridge, UK: Cambridge University Press, 1997); Rebekah Farrugia and Thom Swiss, "Producing Producers: Women and Electronic/Dance Music," *Current Musicology* 86 (2008): 79–99; Georgina Born and Kyle Devine, "Music Technology, Gender, and Class: Digitization, Educational and Social Change in Britain," *Twentieth-Century Music* 12 (2015): 135–172.

74. "Calvin Harris on Dance-Pop as a 'Futuristic Experiment,'" NPR, December 7, 2012, https://www.npr.org/templates/transcript/transcript.php?storyId=166756595.

75. "Calvin Harris Says Creating Pop Music is Easier than Making a Ham Sandwich," *NME*, December 22, 2012, https://www.nme.com/news/music/calvin-harris-44-1248993.

76. "Interview: The Rise of Felix Snow."

77. Cusick, "On Musical Performances of Gender and Sex," 32.

78. Alexis Petridis, "Calvin Harris: Ready for the Weekend," *The Guardian*, August 6, 2009, https://www.theguardian.com/music/2009/aug/07/calvin-harris-cd-review.

79. Marc Hogan, "Calvin Harris: I Created Disco," *Pitchfork*, November 1, 2007, https://pitchfork.com/reviews/albums/10798-i-created-disco/.

80. "Calvin Harris Gives Up Singing Live," BBC, November 30, 2010, http://www.bbc.co.uk/newsbeat/article/11864899/calvin-harris-gives-up-singing-live.

81. Zane Lowe, "Interview with Calvin Harris," Beats 1 Radio on Apple Music, September 19, 2016,.

82. "Calvin Harris on Dance-Pop as a 'Futuristic Experiment.'"

83. Lowe, "Interview with Calvin Harris."

84. See Dana Gooley, *The Virtuoso Liszt* (Cambridge, UK: Cambridge University Press, 2004), esp. 106–109; Philip Auslander, "Lucille Meets GuitarBot: Instrumentality, Agency, and Technology in Musical Performance," *Theatre Journal* 61 (2009): 603–616.

85. Alex Comfort, *The Joy of Sex* (New York: Pocket Books, 2002), 12. The idea can indeed be found in early twentieth-century marriage manuals. As British medical professional and birth control advocate Helena Wright wrote in *The Sex Factor in Marriage: A Book for Those Who Are About to Be Married* (New York: Vanguard, 1932), 86: "A woman's body can be regarded as a musical instrument awaiting the hand of an artist. Clumsiness and ignorance will produce nothing but discord, knowledge and skill evoke responses of limitless beauty." Thank you to Caroline Rusterholz for directing me to this source.

86. "Interview: The Rise of Felix Snow."

87. "What Is Happening at the Beginning of Kiiara's 'Gold.'"

88. Kai Arne Hansen, "Empowered or Objectified? Personal Narrative and Audiovisual Aesthetics in Beyoncé's Partition," *Popular Music and Society* 40 (2017): 171. Hansen also considers a singer's manipulation of his or her voice an aspect of "sonic staging."

89. Brennan Carley, "*SPIN* Pop Report: All Saints' Reunion Unexpectedly Rules, Kiiara's a Cyborg but She's Our Cyborg," *Spin*, June 28, 2016, https://www.spin.com/2016/06/spin-pop-report-all-saints-kiiara-chloe-x-halle-ladyhawke-that-poppy-chelsea-lankes-tiffany/.

90. Richard S. Chang, "MNEK: Diva Hitmaker Steps Out of the Shadows," *Red Bull*, March 19, 2017, https://www.redbull.com/us-en/mnek-beyonce-and-madonna-songwriter-going-solo.

91. "MNEK Reveals How He Writes a Hit Pop Song," BBC News, July 11, 2016, https://www.bbc.com/news/av/entertainment-arts-36747207/mnek-reveals-how-he-writes-a-hit-pop-song.

92. Haraway, "A Manifesto for Cyborgs," 66.

93. Andy Clark, *Natural-Born Cyborgs: Minds, Technologies, and the Future of Human Intelligence* (Oxford: Oxford University Press, 2003).

94. Wynter and McKittrick, "Unparalleled Catastrophe for Our Species?," 36.

95. Katherine Ott, "The Sum of Its Parts: An Introduction to Modern Histories of Prosthetics," in *Artificial Parts, Practical Lives: Modern Histories of Prosthetics*, ed. Katherine Ott, David Serlin, and Stephen Mihm (New York: New York University Press, 2002), 4. Raiford Guins and Omayra Zaragoza Cruz, "Prosthetists at 33 1/3," in *The Prosthetic Impulse: From a Posthuman Present to a Biocultural Future*, ed. Marquard Smith and Joanne Morra (Cambridge, MA: MIT Press, 2005), 228.

CODA

1. Eleanor Sandry, "Creative Collaborations with Machines," *Philosophy of Technology* 30, no. 3 (2017): 311, 309–310.

2. Jayson Greene, "Do Androids Dream of Electric Guitars? Exploring the Future of Musical AI," *Pitchfork*, June 12, 2017, https://pitchfork.com/features/overtones/10091-do-androids-dream-of-electric-guitars-exploring-the-future-of-musical-ai/.

3. Arthur Miller, *The Artist in the Machine* (Cambridge, MA: MIT Press, 2019), 140.

4. Paul Théberge, *Any Sound You Can Imagine: Making Music/Consuming Technology* (Hanover, NH: Wesleyan University Press, 1997).

5. Flow Machines, accessed April 6, 2022, https://www.flow-machines.com. On Flow Machines, see Melissa Avdeef, "Artificial Intelligence & Popular Music: SKYGGE, Flow Machines, and the Audio Uncanny Valley," *Arts* 8, no. 4 (2019): 130, https://doi.org/10.3390/arts8040130.

6. Greene, "Do Androids Dream of Electric Guitars?"

7. George E. Lewis, "Too Many Notes: Complexity and Culture in Voyager," *Leonardo Music Journal* 10 (2000): 34. For a comparison of *Voyager* to other improvising computer systems, see Brian A. Miller, "'All of the Rules of Jazz': Stylistic Models and Algorithmic Creativity in Human-Computer Improvision," *Music Theory Online* 26, no. 3 (2020).

8. "George Lewis and Vijay Iyer in Concert," WellesleyCollege, February 21, 2012, YouTube video, https://youtu.be/IBPJ2HAmsc8. The concert took place February 10, 2012, in Houghton Chapel, Wellesley, Massachusetts.

9. This event was recorded and is available as Laetitia Sonami, "Your Presence Is Required: Performance, Control and Magnetism," posted by Berkeley Center for New Media to Internet Archive, https://archive.org/details/sonami.

10. See Laura Otis, *Networking: Communicating with Bodies and Machines in the Nineteenth Century* (Ann Arbor: University of Michigan Press, 2001).

11. Tara H. Abraham, "Intellectual Origins of the McCulloch-Pitts Neural Networks," *Journal of the History of the Behavioral Sciences* 38 (2002): 3–25.

12. Warren McCulloch and Walter Pitts, "A Logical Calculus of the Ideas Immanent in Nervous Activity," *The Bulletin of Mathematical Biophysics* 5 (1943):115–133.

13. Bohdan Macukow, "Neural Networks—State of Art, Brief History, Basic Models and Architecture," in *Computer Information Systems and Industrial Management*, ed. Khalid Saeed and Władysław Homenda (Cham, Switzerland: Springer, 2016), 3–14.

14. Rebecca Fiebrink and Laetitia Sonami, "Reflections on Eight Years of Instrument Creation with Machine Learning," Proceedings of the International Conference on New Interfaces for Musical Expression, 3, Birmingham City University, 2020.

15. Dadabots, *Coditany of Timeness*, Bandcamp, November 16, 2017, https://dadabots.bandcamp.com/album/coditany-of-timeness.

16. "Listening to Computer-Generated Music Metal," The Needle Drop, January 24, 2018, YouTube video, https://youtu.be/0zOprEiOOyc.

17. CJ Carr and Zack Zukowski, "Generating Albums with SampleRNN to Imitate Metal, Rock, and Punk Bands," *Proceedings of the 6th International Workshop on Musical Metacreation* (MUME 2018), https://doi.org/10.48550/arXiv.1811.06633.

18. Carter Moon, Merry-Go-Roundtable Episode 22: Dadabots Interview, *Merry-Go-Round Magazine*, May 14, 2019, https://soundcloud.com/merrygoroundmagazine/episode-22-dadabots-interview.

19. "We Trolled EUROVISION," Dadabots, April 16, 2020, YouTube video, https://www.youtube.com/watch?v=Sadou56_JUg.

20. CJ Carr, presentation at Northeastern University, November 23, 2020. To give another example—describing her machine learning model trained on Bach chorales in order to harmonize new melodies in that style (deployed in the 2019 Bach Google Doodle), musician and Google Brain researcher scientist Cheng-Zhi Anna Huang compared "interacting with the model" to "learning to play an instrument, because the model has constraints" and there are learnable relationships between the kind of melody one gives it and the kind of harmonization it will produce. Finn Upham, "Music Transformer and Machine Learning for Composition with Dr. Anna Huang," *The So Strangely Podcast*, August 17, 2019, https://www.listennotes.com/podcasts/the-so-strangely/music-transformer-and-Sya26bgfv-M/.

21. "Reeps One ft. A.I. 'Second Self' (We Speak Music—Episode 6—Human and Machine)," Swissbeatbox, April 27, 2019, YouTube video, https://www.youtube.com/watch?v=q981cTdL0_Y.

22. "We Speak Music—Episode 5—The Artificial Artist," Swissbeatbox, March 30, 2019, YouTube video, https://youtu.be/vuwRlTtFzkQ.

23. July Gray, "Holly Herndon x Jlin—'Godmother' (Feat. Spawn) Video," *Stereogum*, December 4, 2018, https://www.stereogum.com/2024764/holly-herndon-x-jlin-godmother-feat-spawn-video/music/.

24. Lars Gotrich, "Holly Herndon's AI Baby Sings to her 'Godmother,'" NPR, December 4, 2018, https://www.npr.org/sections/allsongs/2018/12/04/672758884/holly-herndons-ai-baby-sings-to-her-godmother.

25. On the state of audio style transfer around this time, see Eric Grinstein, Ngoc Q. K. Duong, Alexey Ozerov, and Patrick Peerez, "Audio Style Transfer," 2018 IEEE International Conference on Acoustics, Speech and Signal Processing (ICASSP), revised November 7, 2018, https://arxiv.org/pdf/1710.11385.pdf. Herndon's subsequent project, Holly+, released in 2021, allowed anyone to upload an audio track to a vocal model trained on her voice and to hear the result, which "retains the pitches and rhythms of a user-uploaded audio file, but adds textures and timbres learned from the training set." Holly Herndon, "Holly+," July 13, 2021, https://holly.mirror.xyz/54ds2IiOnvthjGFkokFCoaI4EabytH9xjAYy1irHy94.

26. "The Best Songs of 2015," *New York Times*, December 15, 2015, https://www.nytimes.com/2015/12/15/arts/music/the-best-songs-of-2015.html.

27. Jayson Greene, "Embryo EP," *Pitchfork*, December 10, 2021, https://pitchfork.com/reviews/albums/jlin-embryo-ep/.

28. Peter Kirn, "Jlin, Holly Herndon, and 'Spawn' Find Beauty in AI's Flaws," *CDM*, December 10, 2018, https://cdm.link/2018/12/jlin-holly-herndon-and-spawn-find-beauty-in-ais-flaws/.

29. Andy Beta, "Inside the World's First Mainstream Album Made with AI," *Vulture*, November 13, 2019, https://www.vulture.com/2019/11/holly-herndon-on-proto-an-album-made-with-ai.html.

30. Leah Mandel, "Holly Herndon's New, AI-Spawned Album is Full of Humanity," *Vice*, May 10, 2019, https://www.vice.com/en/article/kzmzxe/holly-herndons-explains-ai-spawn-new-album.

31. "Holly Herndon—Birthing PROTO," Holly Herndon, September 10, 2019, YouTube video, https://www.youtube.com/watch?v=v_4UqpUmMkg.

32. Lewis, "Why Do We Want Our Computers to Improvise?"; Bruno Latour, "Love Your Monsters," *The Breakthrough*, February 14, 2012, https://thebreakthrough.org/journal/issue-2/love-your-monsters; Edward Lewis, Noelani Arista, Archer Pechawis, and Suzanne Kite, "Kin with the Machines," *Journal of Design and Science* (2018), https://doi.org/10.21428/bfafd97b.

33. The music video, posted on YouTube on December 4, 2018, was titled "Holly Herndon & Jlin (feat. Spawn)—Godmother (Official Video)," https://www.youtube.com/watch?v=sc9OjL6Mjq0. Apple Music also included "(feat. Spawn)" in the track name, while Spotify provided simply "Godmother" as the track name and listed Holly Herndon, Jlin, and Spawn as the artists.

34. These included Endel (erroneously reported to be the first algorithm signed to a record deal, rather than a distribution deal between the company Endel and Warner Music) and AIVA (an AI tool designed to compose film scores, as well as the name registered as a composer with the authors' rights society of France and Luxembourgh, SACEM). Dani Deahl, "Warner Music Signed an Algorithm to a Record Deal—What Happens Next?," *The Verge*, March 27, 2019, https://www.theverge.com/2019/3/27/18283084/warner-music-algorithm-signed-ambient-music-endel; Ed Lauder, "Aiva is the first AI to Officially be Recognized as a Composer," *AI Business*, March 10, 2017, https://aibusiness.com/verticals/aiva-is-the-first-ai-to-officially-be-recognised-as-a-composer.

35. The album *PROTO* traces the metaphorical life course of Spawn. The album begins with "Birth," ends with "Last Gasp," and in between includes "live training" sessions that offer a small window onto the data from which Spawn learned (one live training track features a cappella singing, the other a call-and-response between singing ensemble and Spawn's emulation).

36. Jordan Darville, "Holly Herndon on the Power of Machine Learning and Developing Her 'Digital Twin' Holly+," *The Fader*, July 21, 2021, https://www.thefader.com/2021/07/27/holly-herndon-on-the-power-of-machine-learning-and-developing-her-digital-twin-holly.

37. Memo Akten, "Ultrachunk (2018)," https://www.memo.tv/works/ultrachunk/.

38. Barry McHugh, "Making History: Jennifer Walshe," *Painting in Text*, October 18, 2018, https://paintingintext.com/2018/10/18/making-history-jennifer-walshe/.

39. Hexorcismos, *Transfiguración*, Bandcamp, May 1, 2020, https://hexorcismos.bandcamp.com/album/.

40. "Bronze," Bronze, accessed April 16, 2022, https://bronze.ai/.

41. "Dreams of a Machine," MoMA, *Magazine*, July 14, 2020, https://www.moma.org/magazine/articles/378.

42. "Dreams of a Machine."

43. Arca press release quoted in Madison Bloom, "Arca Shares 100 New Versions of 'Riquiquí': Listen," *Pitchfork*, December 16, 2020, https://pitchfork.com/news/arca-shares-100-new-versions-of-riquiqui-listen/.

44. Matt Moen, "Arca: Embracing the Flux," *PaperMag*, April 7, 2020, https://www.papermag.com/arca-transformation-2645630264.html.

45. Alex Frank, "Arca Is the Future We Hope For," *Garage Magazine* 18 (March 8, 2020), https://garage.vice.com/en_us/article/bvgp75/arca-angel; Wren Sanders, "Now List 2020: The Divine Mutability of Arca," *Them* (June 26, 2020), https://www.them.us/story/now-list-2020-arca-interview.

46. For useful overviews, see Declan McGlynn, "AI Futures: How Artificial Intelligence Will Shape Music Production," *DJ Mag*, October 6, 2021, https://djmag.com/longreads/ai-futures-how-artificial-intelligence-will-shape-music-production; Dorien Herremans, Ching-Hua Chuan, and Elain Chew, "A Function Taxonomy of Music Generation Systems," *ACM Computing Surveys* 50, no. 5 (September 2018): 1–30, https://doi.org/10.1145/3108242.

47. Jesse Engle, Lamtharn Hantrakul, Chenjie Gu, Adam Roberts, "DDSP: Differentiable Digital Signal Processing," International Conference on Learning Representations (ICLR) 2020, https://openreview.net/pdf?id=B1x1ma4tDr (on their selection of violin data set, from performances by John Garner, see p. 7); "DDSP: Differentiable Digital Signal Processing: Online Supplement," https://storage.googleapis.com/ddsp/index.html. The vocal performance in the audio example under discussion here is by musician and research team member Lamtharn (Hanoi) Hantrakul, singing the beginning of "Somewhere Over the Rainbow."

48. "Approachable AI for Music, Model Markets, New DAWs and Holly+ with Never Before Heard Sounds," *Interdependence* podcast, July 14, 2021, https://interdependence.simplecast.com/episodes/approachable-ai-for-music-model-markets-new-daws-and-holly-with-never-before-heard-sounds-OBSZOk77.

49. For a perspective on hands as sites of embodied knowledge and hope for the future, consider the significance of carving a doll from wood at the end of Sylvia Wynter's novel *The Hills of Hebron* (Kingston, Jamaica: Ian Randle Publishers, [1962] 2010). For a perspective on the "by hand" positioned in hopeless battle for humanity against encroaching algorithmic automation, see the discussion of competing approaches to music recommendation in Nick Seaver, *Computing Taste: Algorithms and the Makers of Music Recommendation* (Chicago: University of Chicago Press, 2022). See also Jonathan de Souza, *Music at Hand: Instruments, Bodies, and Cognition* (Oxford: University of Oxford Press, 2017).

50. Jack Malcolm, "Robots Can Make Music, but Can They Sing?," *New York Times*, July 7, 2021.

Works Cited

ARCHIVAL COLLECTIONS AND FREQUENTLY CITED PERIODICALS

The Daphne Oram Archive, Special Collections & Archives, Goldsmiths University of London
Mémoires de l'Académie Royale des Sciences
Mercure de France
Pitchfork
The New York Times
The Village Voice
The Washington Post
The Wire

BOOKS AND ARTICLES

Abbate, Carolyn. *In Search of Opera*. Princeton, NJ: Princeton University Press, 2001.
Abbate, Carolyn. "Outside Ravel's Tomb." *Journal of the American Musicological Society* 52, no. 3 (1999): 465–530.
Abraham, Tara H. "Intellectual Origins of the McCulloch-Pitts Neural Networks." *Journal of the History of the Behavioral Sciences* 38 (2002): 3–25.
Alano, Jomarie. "The Triumph of the *bouffons*: La serva padrona at the Paris Opera, 1752–1754." *The French Review* 79, no. 1 (2005): 124–135.
Allanbrook, Wye Jamison. *Comic Mimesis in Late Eighteenth-Century Music*, edited by Mary Ann Smart and Richard Taruskin. Oakland: University of California Press, 2014.
Anthony, James R. *French Baroque Music from Beaujoyeulx to Rameau*. Oxford: Oxford University Press, 1973.
"Approachable AI for Music, Model Markets, New DAWs and Holly+ with Never Before Heard Sounds." *Interdependence* podcast, July 14, 2021. https://interdependence.simplecast.com/episodes/approachable-ai-for-music-model-markets-new-daws-and-holly-with-never-before-heard-sounds-OBSZOk77.
Auner, Joseph. "Losing Your Voice: Sampled Speech and Song from the Uncanny to the Unremarkable." In *Throughout: Art and Culture Emerging with Ubiquitous Computing*, edited by Ulrik Ekman, 135–149. Cambridge, MA: MIT Press, 2013.

Auner, Joseph. "'Sing It for Me': Posthuman Ventriloquism in Recent Popular Music." *Journal of the Royal Musical Association* 128 (2003): 98–122.

Auslander, Philip. "Fluxus Art-Amusement: The Music of the Future?" In *Contours of the Theatrical Avant-Garde: Performance and Textuality*, edited by James M. Harding, 110–129. Ann Arbor: University of Michigan Press, 2000.

Auslander, Philip. "Lucille Meets GuitarBot: Instrumentality, Agency, and Technology in Musical Performance." *Theatre Journal* 61 (2009): 603–616.

Avdeef, Melissa. "Artificial Intelligence & Popular Music: SKYGGE, Flow Machines, and the Audio Uncanny Valley." *Arts* 8, no. 4 (2019): 130. https://doi.org/10.3390/arts8040130.

Bacon, Francis. *The New Organon*, edited by L. Jardin and M. Silverthorn. Cambridge Texts in the History of Philosophy. Cambridge, UK: Cambridge University Press, 2000.

Banerji, Ritwik. "Whiteness as Improvisation, Nonwhiteness as Machine." *Jazz & Culture* 4, no. 2 (2021): 56–84.

Barzun, Jacques. "Introductory Remarks to a Program of Works Produced at the Columbia-Princeton Electronic Music Center." In *Audio Culture: Readings in Modern Music*, edited by Christoph Cox and Daniel Warner, 367–369. New York: Bloomsbury Academic, 2004.

Bates, Eliot. "The Social Life of Musical Instruments." *Ethnomusicology* 56, no. 3 (2012): 363–395.

Baumann, Thomas. *North German Opera in the Age of Goethe*. Cambridge, UK: Cambridge University Press, 1985.

Beals, Kurt. "'Do the New Poets Think? It's Possible': Computer Poetry and Cyborg Subjectivity." *Configurations* 26, no. 2 (2018): 149–177.

Beecher, Donald. "Aesthetics of the French Solo Viol Repertory, 1650–1680." *Journal of the Viola da Gamba Society of America* 24 (1987): 10–21.

Belfi, Amy M., Anna Kasdan, and Daniel Tranel. "Anomia for Musical Entities." *Aphasiology* 44, no. 3 (2019): 382–404.

Benhamou, Reed. "From Curiosité to Utilité: The Automaton in Eighteenth-Century France." *Studies in Eighteenth-Century Culture* 17 (1988): 91–105.

Bennett, Jane. *Vibrant Matter: A Political Ecology of Things*. Durham, NC: Duke University Press, 2010.

Berland, Jody. "The Musicking Machine." In *Residual Media*, edited by Charles Acland, 155–184. Minneapolis: University of Minnesota Press, 2007.

Beta, Andy. "Inside the World's First Mainstream Album Made with AI." *Vulture*, November 13, 2019. https://www.vulture.com/2019/11/holly-herndon-on-proto-an-album-made-with-ai.html.

Blacking, John. *How Musical Is Man?* Seattle: University of Washington Press, 1974.

Bloechl, Olivia. *Native American Song at the Frontiers of Early Modern Music*. Cambridge, UK: Cambridge University Press, 2008.

Born, Georgina, and Kyle Devine. "Music Technology, Gender, and Class: Digitization, Educational and Social Change in Britain." *Twentieth-Century Music* 12 (2015): 135–172.

Borthwick, E. K. "The Riddle of the Tortoise and the Lyre." *Music and Letters* 51, no. 4 (1970): 373–387.

Boscagli, Maurizia. *Stuff Theory: Everyday Objects, Radical Materialism.* New York: Bloomsbury, 2014.
Bowers, Jane. "A Catalogue of French Works for the Transverse Flute 1692–1761." *Recherches sur la musique française classique* 28 (1978): 89–125.
Brown, Bill. "Thing Theory." *Critical Inquiry* 28, no. 1 (2001): 1–22.
Buch, David. *Magic Flutes & Enchanted Forests: The Supernatural in Eighteenth-Century Musical Theater.* Chicago: University of Chicago Press, 2008.
Burton, Justin. *Posthuman Rap.* Oxford: Oxford University Press, 2017.
Bussy, Pascal. *Kraftwerk: Man Machine and Music*, 3rd ed. London: SAF Publishing, 2005.
Carr, CJ, and Zack Zukowski. "Generating Albums with SampleRNN to Imitate Metal, Rock, and Punk Bands." *Proceedings of the 6th International Workshop on Musical Metacreation* (MUME 2018). https://doi.org/10.48550/arXiv.1811.06633.
Carr, J. L. "Pygmalion and the Philosophes: The Animated Statue in Eighteenth-Century France." *Journal of the Warburg and Courtauld Institutes* 23, no. 3-4 (1960): 239–255.
Cateforis, Theo. *Are We Not New Wave?: Modern Pop at the Turn of the 1980s.* Ann Arbor: University of Michigan Press, 2011.
Chadabe, Joel. *Electric Sound: The Past and Promise of Electronic Music.* Upper Saddle River, NJ: Prentice Hall, 1997.
Christensen, Thomas. *Rameau and Musical Thought in the Enlightenment.* Cambridge, UK: Cambridge University Press, 1993.
Chua, Daniel. *Absolute Music and the Construction of Meaning.* Cambridge, UK: Cambridge University Press, 1999.
Chude Sokei, Louis. *The Sound of Culture: Diaspora and Black Technopoetics.* Middletown, CT: Wesleyan University Press, 2015.
Chun, Wendy Hui Kyong. "Introduction: Race and/as Technology; or, How to Do Things with Race." *Camera Obscura* 24, no. 1 (2009): 7–35.
Clark, Andy. *Natural-Born Cyborgs: Minds, Technologies, and the Future of Human Intelligence.* Oxford: Oxford University Press, 2003.
Cohen, David E. "The 'Gift of Nature': Musical 'Instinct' and Musical Cognition in Rameau." In *Music Theory and Natural Order from the Renaissance to the Early Twentieth Century*, edited by Suzannah Clark and Alexander Rehding, 68–92. Cambridge, UK: Cambridge University Press, 2001.
Coleman, Beth. "Race as Technology." *Camera Obscura* 24, no. 1 (2009): 177–207.
Cook, Susan C., and Judy S. Tsou, eds. *Cecilia Reclaimed: Feminist Perspectives on Gender and Music.* Chicago: University of Illinois Press, 1994.
Cope, David. *Virtual Music: Computer Synthesis of Musical Style.* Cambridge, MA: MIT Press, 2001.
Cottom, Daniel. "Art in the Age of Mechanical Digestion." *Representations* 66 (1999): 52–74.
Cowart, Georgia. *The Triumph of Pleasure: Louis XIV & the Politics of Spectacle.* Chicago: University of Chicago Press, 2008.
Cowart, Georgia. "Watteau's *Pilgrimage to Cythera* and the Subversive Utopia of the Opera-Ballet." *Art Bulletin* 83, no. 3 (2001): 461–478.

Cusick, Suzanne. "On Musical Performances of Gender and Sex." In *Audible Traces: Gender, Identity, and Music*, edited by Elaine Barkin and Lydia Hanessley, 25–48. Los Angeles: Carciofoli, 1999.

Cypess, Rebecca. *Curious and Modern Inventions: Instrumental Music as Discovery in Galileo's Italy*. Chicago: University of Chicago Press, 2016.

Cypess, Rebecca, and Steven Kemper. "The Anthropomorphic Analogy: Humanising Musical Machines in the Early Modern and Contemporary Eras." *Organised Sound* 23, no. 2 (2018): 167–180.

D'Argenville, Antoine Nicolas Dézallier. *Voyage pittoresque des environs de Paris*. Paris: De Bure, 1752.

Darnton, Robert. *The Literary Underground of the Old Regime*. Cambridge, MA: Harvard University Press, 1982.

Darville, Jordan. "Holly Herndon on the Power of Machine Learning and Developing Her 'Digital Twin' Holly+." *The Fader*, July 21, 2021. https://www.thefader.com/2021/07/27/holly-herndon-on-the-power-of-machine-learning-and-developing-her-digital-twin-holly.

Daston, Lorraine, and Katherine Park. *Wonders and the Order of Nature, 1150–750*. New York: Zone Books, 1998.

Davies, James Q. *Romantic Anatomies of Performance*. Berkeley: University of California Press, 2014.

Davies, Stephen. "What Is the Sound of One Piano Plummeting?" In *Themes in the Philosophy of Music*. Oxford: Oxford University Press, 2003.

Dery, Mark. "Black to the Future: Interviews with Samuel R. Delany, Greg Tate, and Tricia Rose." *South Atlantic Quarterly* 94 (1993): 735–788.

Des Chene, Dennis. *Spirits and Clocks: Machine and Organism in Descartes*. Ithaca, NY: Cornell University Press, 2001.

De Souza, Jonathan. *Music at Hand: Instruments, Bodies, and Cognition*. Oxford: University of Oxford Press, 2017.

De Souza, Jonathan. "Voice and Instrument at the Origins of Music." *Current Musicology* 97 (2014): 21–36.

Dickinson, Kay. "'Believe'? Vocoders, Digitalised Female Identity and Camp." *Popular Music* 20 (2001): 333–347.

Diderot, Denis. *Rameau's Nephew: Le Neveu de Rameau*. Edited by Marian Hobson. Translated by Kate E. Tunstall and Caroline Warman. Cambridge, UK: Open Book Publishers, 2016.

Dill, Charles. *Monstrous Opera: Rameau and the Tragic Tradition*. Princeton, NJ: Princeton University Press, 1998.

Dixon, Mike J., Marie Piskopos, and Tom A. Schweizer. "Musical Instrument Naming Impairments: The Crucial Exception to the Living/Nonliving Dichotomy in Category-Specific Agnosia." *Brain and Cognition* 43 (2000): 158–164.

Dolan, Emily I. *The Orchestral Revolution: Haydn and the Technologies of Timbre*. Cambridge, UK: Cambridge University Press, 2013.

Dolan, Emily I., and Thomas Patteson. "Ethereal Timbres." In *The Oxford Handbook of Timbre*, edited by Emily I. Dolan and Alexander Rehding, 183–204. Oxford: Oxford University Press, 2021.

Donaheu, Charles. "The Legal Background: European Marriage Law from the

Sixteenth to the Nineteenth Century." In *Marriage in Europe, 1400–1800*, edited by Silvana Seidel Menchi, 33–60. Toronto: University of Toronto Press, 2016.

Doyon, André, and Lucien Liaigre. *Jacques Vaucanson, mécanicien de genie*. Paris: Presses universitaires de France, 1967.

Eidsheim, Nina. "'Sonic Blackness' in American Opera." *American Quarterly* 63 (2011): 641–671.

Ellis, Katharine. "Female Pianists and Their Male Critics in Nineteenth-Century Paris." *Journal of the American Musicological Society* 50 (1997): 353–385.

Engle, Jesse, Lamtharn Hantrakul, Chenjie Gu, and Adam Roberts. "DDSP: Differentiable Digital Signal Processing." International Conference on Learning Representations (ICLR), 2020. https://openreview.net/pdf?id=B1x1ma4tDr.

Erlmann, Veit. *Reason and Resonance: A History of Modern Aurality*. New York: Zone Books, 2010.

Eshun, Kodwo. *More Brilliant Than the Sun: Adventures in Sonic Fiction*. London: Quartet Books, 1998.

Farrugia, Rebekah, and Thom Swiss. "Producing Producers: Women and Electronic/Dance Music." *Current Musicology* 86 (2008): 79–99.

Fermel'huis, Jean-Baptiste. *Eloge funèbre Monsieur Coysevox*. Paris: Collombat, 1721.

Festa, Lynn. *Fiction without Humanity: Person, Animal, Thing in Early Enlightenment Literature and Culture*. Philadelphia: University of Pennsylvania Press, 2019.

Festa, Lynn. *Sentimental Figures of Empire in Eighteenth-Century Britain and France*. Baltimore, MD: Johns Hopkins University Press, 2006.

Fiebrink, Rebecca, and Laetitia Sonami. "Reflections on Eight Years of Instrument Creation with Machine Learning." *Proceedings of the International Conference on New Interfaces for Musical Expression*, 237–242. Birmingham City University, 2020. http://doi.org/10.5281/zenodo.4813334.

Frank, Alex. "Arca Is the Future We Hope For." *Garage Magazine* 18 (March 8, 2020). https://garage.vlce.com/en_us/article/bvgp75/arca-angel.

Gengaro, Christine Lee. *Listening to Stanley Kubrick: The Music in His Films*. Toronto: Scarecrow Press, 2013.

Gerovitch, Slava. *From Newspeak to Cyberspeak: A History of Soviet Cybernetics*. Cambridge, MA: MIT Press, 2002.

Goehr, Lydia. *The Quest for Voice*. Oxford: Oxford University Press, 2002.

Goodwin, Andrew. "Sample and Hold: Pop Music in the Digital Age of Reproduction." *Critical Quarterly* 30, no. 3 (1988): 34–49.

Gooley, Dana. *The Virtuoso Liszt*. Cambridge, UK: Cambridge University Press, 2004.

Gordon, Bonnie. "L'Orfeo at 400: Orfeo's Machines." *Opera Quarterly* 24, no. 3–4 (2009): 200–222.

Gotrich, Lars. "Holly Herndon's AI Baby Sings to her 'Godmother.'" NPR, December 4, 2018. https://www.npr.org/sections/allsongs/2018/12/04/672758884/holly-herndons-ai-baby-sings-to-her-godmother.

Gouk, Penelope. "Clockwork or Musical Instrument? Some English Theories of Mind-Body Interaction before and after Descartes." In *Structures of Feeling in Seventeenth-Century Cultural Expression*, edited by Susan McClary, 35–59. Toronto: University of Toronto Press, 2013.

Graffigny, Françoise de. *Letters of a Peruvian Woman*. Translated by Jonathan Mallinson. Oxford: Oxford University Press, 2009.

Grant, Roger. "Editorial." *Eighteenth-Century Music* 17, no. 1 (2020): 5–8.

Grant, Roger. "Peculiar Attunements: Comic Opera and Enlightenment Mimesis." *Critical Inquiry* 43 (2017): 550–569.

Gray, July. "Holly Herndon x Jlin—'Godmother' (Feat. Spawn) Video." *Stereogum*, December 4, 2018. https://www.stereogum.com/2024764/holly-herndon-x-jlin-godmother-feat-spawn-video/music/.

Green, Lucy. *Music, Gender, Education*. Cambridge, UK: Cambridge University Press, 1997.

Green, Robert A. "Jean Rousseau and Ornamentation in French Viol Music." *Journal of the Viola da Gamba Society of America* 14 (1977): 4–41.

Grinstein, Eric, Ngoc Q. K. Duong, Alexey Ozerov, and Patrick Peerez. "Audio Style Transfer." 2018 IEEE International Conference on Acoustics, Speech and Signal Processing (ICASSP), revised November 7, 2018. https://arxiv.org/pdf/1710.11385.pdf.

Guins, Raiford, and Omayra Zaragoza Cruz. "Prosthetists at 33 1/3." In *The Prosthetic Impulse: From a Posthuman Present to a Biocultural Future*, edited by Marquard Smith and Joanne Morra, 221–236. Cambridge, MA: MIT Press, 2005.

Hadlock, Heather. *Mad Loves: Women and Music in Offenbach's Les contes d'Hoffman*. Princeton, NJ: Princeton University Press, 2000.

Hamilton, Jack. *Just Around Midnight*. Cambridge, MA: Harvard University Press, 2016.

Hansen, Kai Arne. "Empowered or Objectified? Personal Narrative and Audiovisual Aesthetics in Beyoncé's Partition." *Popular Music and Society* 40, no. 2 (2017): 164–180.

Haraway, Donna. "A Manifesto for Cyborgs: Science, Technology, and Socialist Feminism in the 1980s." *Socialist Review* 15, no. 2 (1985): 65–107.

Hayes, Deborah. "Marie-Emmanuelle Bayon, Later Madame Louis, and Music in Late Eighteenth-Century France." *College Music Symposium* 30, no. 1 (1990): 14–33.

Hayles, N. Katherine. *How We Became Posthuman: Virtual Bodies in Cybernetics, Literature, and Informatics*. Chicago: University of Chicago Press, 1999.

Hayles, N. Katherine. "Self-Reflexive Metaphors in Maxwell's Demon and Shannon's Choice: Finding the Passages." In *Literature and Science: Theory & Practiced*, edited by Stuart Peterfreund, 209–237. Boston: Northeastern University Press, 1990.

Heller, Wendy. "Dancing Statues and the Myth of Venice: Ancient Sculpture on the Opera Stage." *Art History* 33, no. 2 (2010): 304–319.

Herremans, Dorien, Ching-Hua Chuan, and Elain Chew. "A Function Taxonomy of Music Generation Systems." *ACM Computing Surveys* 50, no. 5 (September 2018): 1–30. https://doi.org/10.1145/3108242.

Herzog, Myrna. "Is the Quinton a Viol? A Puzzle Unravelled," *Early Music* 28, no. 1 (2000): 8–31.

Hirt, Katherine. *When Machines Play Chopin: Musical Spirit and Automation in Nineteenth-Century German Literature*. New York: Walter de Gruyter, 2010.

Hollingworth, Roy. "The Walter Carlos Sonic Boom." *Melody Maker*, September 23, 1972. https://www.rocksbackpages.com/Library/Article/the-walter-carlos-sonic-boom.
Jackson, Barbara Garvey. "Hubert Le Blanc's *Defense de la viole*." *Journal of the Viola da Gamba Society of America* 10 (1973): 11–28, 69–80; 11 (1974), 17–58.
Jackson, Zakiyyah Iman. *Becoming Human: Matter and Meaning in an Antiblack World*. New York: NYU Press, 2020.
James, Robin. "Robo-Diva R&B: Aesthetics, Politics, and Black Female Robots in Contemporary Popular Music." *Journal of Popular Music Studies* 20 (2008): 402–423.
Johnson, Barbara. *Persons and Things*. Cambridge, MA: Harvard University Press, 2008.
Johnson, Edmond. "Policies Intended to Protect Elephants May Inadvertently Endanger Musical Instruments." *Newsletter of the American Musical Instrument Society* 43, no. 2 (2014): 1, 8–10.
Johnson, James H. *Listening in Paris: A Cultural History*. Berkeley: University of California Press, 1995.
Jones, Alisha Lola. *Flaming? The Peculiar Theopolitics of Fire and Desire in Black Male Gospel Performance*. Oxford: Oxford University Press, 2020.
Kallberg, Jeffrey. "Mechanical Chopin." *Common Knowledge* 17, no. 2 (2011): 269–282.
Kang, Minsoo. *Sublime Dreams of Living Machines: The Automaton in the European Imagination*. Cambridge, MA: Harvard University Press, 2011.
Kant, Immanuel. "Was ist Aufklärung?" Translated by Lewis White Beck. In *The Politics of Truth*, edited by Sylvère Lotringer and Lysa Hochroth. New York: Semiotext(e), 2007.
Kass, Lily. "'When then will the veil be listed?': How Translations Obscure Racism in Productions of *The Magic Flute*." Paper presented at American Musicological Society annual meeting, November 2021.
Kassler, Jamie C. "Man—A Musical Instrument: Models of the Brain and Mental Functioning Before the Computer." *History of Science* xxii (1984): 59–92.
Keiser, Jess. *Nervous Fictions: Literary Form and the Enlightenment Origins of Neuroscience*. Charlottesville: University of Virginia Press, 2020.
Kelly, Alexandra Celia. *Consuming Ivory: Mercantile Legacies of East Africa and New England*. Seattle: University of Washington Press, 2021.
Kheshti, Roshanak. *Switched-On Bach*. New York: Bloomsbury Press, 2019.
Kieffer, Alexandra. "Bells and the Problem of Realism in Ravel's Early Piano Music." *Journal of Musicology* 34, no. 3 (2017): 432–472.
Kinney, Gordon J. "A 'Tempest in a Glass of Water' or a Conflict of Esthetic Attitudes." *Journal of the Viola da Gamba Society of America* 14 (1977): 42–52.
Kinney, Gordon J. "Trichet's Treatise: A 17th Century Description of the Viols." *Journal of the Viola da Gamba Society of America* 2 (1965): 16–20.
Kirn, Peter. "Jlin, Holly Herndon, and 'Spawn' Find Beauty in AI's Flaws." *CDM*, December 10, 2018. https://cdm.link/2018/12/jlin-holly-herndon-and-spawn-find-beauty-in-ais-flaws/.
Kline, Robert. "Where are the Cyborgs in Cybernetics?" *Social Studies of Science* 39, no. 3 (2009): 331–362.

Koestenbaum, Wayne. *The Queen's Throat: Opera, Homosexuality and the Mystery of Desire*. New York: Poseidon Press, 1993.

Kovaciny, Stephen M. "Chabanon, the Listening Self and the Prosopopoeia of Aesthetic Experience." *Eighteenth-Century Music* 19, no. 1 (2022): 13–36.

Kreuzer, Gundula. *Curtain, Gong, Steam: Wagnerian Technologies of Nineteenth-Century Opera*. Oakland: University of California Press, 2018.

Kroeger, Brooke. *Passing: When People Can't Be Who They Are*. New York: Public Affairs, 2003.

Lalonde, Amanda. "The Music of the Living-Dead." *Music and Letters* 96 (2015): 602–629.

La Mettrie, Julien Offray de. *Machine Man and Other Writings*. Edited by Ann Thomson. Cambridge, UK: Cambridge University Press, 1996.

Latour, Bruno. "Love Your Monsters." *Breakthrough*, February 14, 2012. https://thebreakthrough.org/journal/issue-2/love-your-monsters.

Latour, Bruno. *We Have Never Been Modern*. Translated by Catherine Porter. Cambridge, MA: Harvard University Press, 1993.

Le Blanc, Hubert. *Defense de la basse de viole contre les entreprises du violon et les pretentions du violoncel*. Amsterdam: Pierre Mortier, 1740.

LeClerc, Hélène. "*Les Indes galantes* (1735–1952): les sources de l'opéra-ballet; l'exotisme orientalisant; les conditions matérielles du spectacles; fortune des *Indes galantes*." *Revue d'histoire du théâtre Paris* 5, no. 4 (1953): 259–285.

Lenhoff, Sylvia G., and Howard M. Lenhoff. *Hydra and the Birth of Experimental Biology—1744: Abraham Trembley's Mémoires concerning the Polyps*. Pacific Grove, CA: Boxwood Press, 1986.

Leppert, Richard. "Sexual Identity, Death, and the Family Piano." *19th-Century Music* 16, no. 2 (1992): 105–128.

Lewin, David. "Women's Voices and the Fundamental Bass." *Journal of Musicology* 10, no. 4 (1992): 464–482.

Lewis, Edward, Noelani Arista, Archer Pechawis, and Suzanne Kite. "Kin with the Machines." *Journal of Design and Science* (2018). https://doi.org/10.21428/bfafd97b.

Lewis, George E. "Too Many Notes: Complexity and Culture in Voyager." *Leonardo Music Journal* 10 (2000): 33–39.

Lewis, George E. "Why Do We Want Our Computers to Improvise?" In *Oxford Handbook of Algorithmic Music*. Edited by Alex McLean and Roger T. Dan. New York: Oxford University Press, 2018.

Lewis, George E. "Why Do We Want Our Computers to Improvise?" Public lecture presented at Monash University, August 13, 2013. https://vimeo.com/78692461.

Licklider, J. C. R. "Man–Computer Symbiosis." *IRE Transactions on Human Factors in Electronics* HFE-1 (March 1960): 4–11.

Liu, Catherine. *Copying Machines: Taking Notes for the Automaton*. Minneapolis: University of Minnesota Press, 2000.

Lloyd, Henry Martyn, ed. *The Discourse of Sensibility: The Knowing Body in the Enlightenment*. New York: Springer, 2013.

Lockhart, Ellen. *Animation, Plasticity and Music in Italy, 1770–1830*. Oakland: University of California Press, 2017.

Lott, R. Allen. *From Paris to Peoria: How European Piano Virtuosos Brought Classical Music to the American Heartland*. Oxford: Oxford University Press, 2003.
Loughridge, Deirdre. "'Always Already Technological': New Views of Music and the Human in Musicology and the Cognitive Sciences." *Music Research Annual* 2 (2021): 1–22. ISSN 2563-7290.
Loughridge, Deirdre. "Who Measured the Wind and Made the Figures Move." *Journal of the American Musicological Society* 66, no. 1 (2013): 270–275.
Lovink, Geer. "An Interview with Kodwo Eshun by Geer Lovink—Originally from 2000, Republished by Blowup Reader 7 (2013)." V2_, Lab for the Unstable Media, Rotterdam, 2013. https://v2.nl/archive/articles/an-interview-with-kodwo-eshun.
Loza, Susana. "Sampling (Hetero)sexuality: Diva-ness and Discipline in Electronic Dance Music." *Popular Music* 20, no. 3 (2001): 349–357.
Macukow, Bohdan. "Neural Networks—State of Art, Brief History, Basic Models and Architecture." In *Computer Information Systems and Industrial Management*, edited by Khalid Saeed and Władysław Homenda, 3–14. Cham, Switzerland: Springer, 2016.
Malcolm, Jack. "Robots Can Make Music, but Can They Sing?" *New York Times*, July 7, 2021.
Mandel, Leah. "Holly Herndon's New, AI-Spawned Album Is Full of Humanity." *Vice*, May 10, 2019. https://www.vice.com/en/article/kzmzxe/holly-herndons-explains-ai-spawn-new-album.
Manning, Céline. "Singer-Machines: Describing Italian Singers, 1800–1850." Translated by Nicholas Manning. *Opera Quarterly* 28 (2012): 230–258.
Maxham, Robert Eugene. "The Contributions of Joseph Sauveur (1653–1716) to Acoustics." PhD diss., University of Rochester, Eastman School of Music, 1976.
Mayhew, Emma. "Positioning the Producer: Gender Divisions in Creative Labour and Value." In *Music, Space, and Place*, edited by Sheila Whiteley, Andy Bennett, and Stan Hawkins, 149–162. Aldershot, UK: Ashgate, 2004.
McClary, Susan. *Feminine Endings*. Minneapolis: University of Minnesota Press, 1991.
McCracken, Allison. *Real Men Don't Sing: Crooning in American Culture*. Durham, NC: Duke University Press, 2015.
McEachern, Jo-Ann, and David Smith. "Mme de Graffigny's *Lettres d'une Péruvienne*: Identifying the First Edition." *Eighteenth-Century Fiction* 9, no. 1 (1996): 21–35.
McGeary, Thomas. "Harpsichord Mottoes." *Journal of the American Musical Instrument Society* 7 (1981): 5–35.
McGlynn, Declan. "AI Futures: How Artificial Intelligence Will Shape Music Production." *DJ Mag*, October 6, 2021. https://djmag.com/longreads/ai-futures-how-artificial-intelligence-will-shape-music-production.
McHugh, Barry. "Making History: Jennifer Walshe." *Painting in Text*, October 18, 2018. https://paintingintext.com/2018/10/18/making-history-jennifer-walshe/.
McLeod, Ken. "Living in the Immaterial World: Holograms and Spirituality in Recent Popular Music." *Popular Music and Society* 39, no. 5 (2016): 501–515.
Meglin, Joellen. "'Sauvages, Sex Roles, and Semiotics': Representations of Native Americans in the French Ballet, 1736–1837, Part One." *Dance Chronicle* 23, no. 2 (2000): 87–132.
Miller, Arthur. *The Artist in the Machine*. Cambridge, MA: MIT Press, 2019.

Miller, Brian A. "'All of the Rules of Jazz': Stylistic Models and Algorithmic Creativity in Human-Computer Improvision." *Music Theory Online* 26, no. 3 (2020).

Mills, Mara. "Media and Prosthesis: The Vocoder, the Artificial Larynx, and the History of Signal Processing." *Qui Parle* 21, no. 1 (2012): 107–149.

Moen, Matt. "Arca: Embracing the Flux." *PaperMag*, April 7, 2020. https://www.papermag.com/arca-transformation-2645630264.html.

Mori, Masahiro. "The Uncanny Valley: The Original Essay by Masahiro Mori." Translated by Karl F. MacDorman and Norri Kageki. *IEEE Spectrum*, June 12, 2012. https://spectrum.ieee.org/the-uncanny-valley.

Morrison, Matthew. "Race, Blacksound, and the (Re)Making of Musicological Discourse." *Journal of the American Musicological Society* 72, no. 3 (2019): 781–823.

Morton, Timothy. *Realist Magic: Objects, Ontology, Causality*. Ann Arbor: Open Humanities Press, 2013.

Moseley, Roger. "Chopin's Aliases." *Nineteenth-Century Music* 42, no. 1 (2018): 3–29.

Moseley, Roger. *Keys to Play: Music as a Ludic Medium from Apollo to Nintendo*. Oakland: University of California Press, 2016.

Mulvey, Laura. *Death 24x a Second: Stillness and the Moving Image*. Chicago: University of Chicago Press, 2006.

Murchison, Gayle. "Let's Flip It! Quare Emancipations: Black Queer Traditions, Afrofuturisms, Janelle Monáe to Labelle." *Women and Music: A Journal of Gender and Culture* 22 (2018): 79–90.

Murray, Sean. "Pianos, Ivory, and Empire." *American Music Review* 38, no. 2 (Spring 2009): 1, 4–5, 13–14.

Murray, Sidney. "Jean-Baptiste Berard's *L'Art du Chant*: Translation and Commentary." PhD diss., University of Iowa, 1965.

Museum of Modern Art. "Dreams of a Machine." *Magazine*, July 14, 2020. https://www.moma.org/magazine/articles/378.

Muthu, Sankar. *Enlightenment Against Empire*. Princeton, NJ: Princeton University Press, 2003.

Nakamura, Lisa. "Prospects for a Materialist Informatics: An Interview with Donna Haraway." *electronic book review*, August 30, 2003.

Nash, Richard. *Wild Enlightenment: The Borders of Human Identity in the Eighteenth Century*. Charlottesville: University of Virginia Press, 2003.

[Nemeitz, Joachim Christoph]. *Sejour de Paris, oder, Getreue Anleitung*. Frankfurt am Main: Friederich Wilhelm Förster, 1718.

Noakes, Richard. "'Instruments to Lay Hold of Spirits': Technologizing the Bodies of Victorian Spiritualism." In *Bodies/Machines*, edited by Iwan Rhys Morus, 125–164. New York: Berg, 2002.

Ochoa Gautier, Ana María. *Aurality: Listening and Knowledge in Nineteenth-Century Colombia*. Durham, NC: Duke University Press, 2014.

Oram, Daphne. *An Individual Note: Of Music, Sound and Electronics*. London: Galliard, 1972.

Otis, Laura. *Networking: Communicating with Bodies and Machines in the Nineteenth Century*. Ann Arbor: University of Michigan Press, 2001.

Ott, Katherine. "The Sum of Its Parts: An Introduction to Modern Histories of Prosthetics." In *Artificial Parts, Practical Lives: Modern Histories of Prosthetics*, edited by Katherine Ott, David Serlin, and Stephen Mihm, 1–42. New York: New York University Press, 2002.

Owen, Marshall. "Auto-Tune In Situ: Articulations of Voice, Affect, and Artifact in the Recording Studio." PhD diss., Cornell University, 2017.

Parsons, Christopher M. *A Not-So-New World: Empire and Environment*. Philadelphia: University of Pennsylvania Press, 2018.

Patteson, Thomas. *Instruments for New Music: Sound, Technology and Modernism*. Oakland: University of California Press, 2015.

Peraino, Judith. "Synthesizing Difference: The Queer Circuits of Early Synthpop." In *Rethinking Difference in Music Scholarship*, edited by Olivia Bloechl, Melanie Lowe, and Jeffrey Kallberg, 287–313. Cambridge, UK: Cambridge University Press, 2015.

Peritz, Jessica Gabriel. "Orpheus's Civilising Song, or, the Politics of Voice in Late Enlightenment Italy." *Cambridge Opera Journal* 31, no. 2–3 (2020): 129–152.

Piroux, Lorraine. "The Encyclopedist and the Peruvian Princess: The Poetics of Illegibility in French Enlightenment Book Culture." *PMLA: Publications of the Modern Language Association of America* 121, no. 1 (2006): 107–123.

Porcello, Thomas. "The Ethics of Digital Audio-Sampling: Engineers' Discourse." *Popular Music* 10, no. 1 (1991): 69–84.

Powers, Ann. *Good Booty: Love and Sex, Black and White, Body and Soul in American Music*. New York: Dey St., 2017.

Powers, Devon. *Writing The Record: The Village Voice and the Birth of Rock Criticism*. Amherst: University of Massachusetts Press, 2013.

Prior, Nick. "Software Sequencers and Cyborg Singers." *New Formations* 66 (2009): 81–99.

Provenzano, Catherine. "Making Voices: The Gendering of Pitch Correction and the Auto-Tune Effect in Contemporary Pop Music." *Journal of Popular Music Studies* 31, no. 2 (2019): 63–84.

Ratcliffe, Marc J. *The Quest for the Invisible: Microscopy in the Enlightenment*. New York: Ashgate, 2009.

Retman, Sonnet. "Between Rock and a Hard Place: Narrating Nona Hendryx's Inscrutable Career." *Women & Performance: A Journal of Feminist Theory* 16, no. 1 (2006): 107–118.

Richards, Annette. "Automatic Genius: Mozart and the Mechanical Sublime." *Music & Letters* 80 (1999): 366–389.

Richards, Annette. *The Free Fantasia and the Musical Picturesque*. Cambridge, UK: Cambridge University Press, 2001.

Richards, Tim. "Oramics: Precedents, Technology, and Influence." Doctoral thesis, Goldsmiths, University of London, 2018.

Riskin, Jessica. "The Defecating Duck, or, the Ambiguous Origins of Artificial Life." *Critical Inquiry* 29, no. 4 (2003): 599–633.

Riskin, Jessica. "Eighteenth-Century Wetware." *Representations* 83, no. 1 (2003): 97–125.

Riskin, Jessica. *The Restless Clock: A History of the Centuries-Long Argument Over What Makes Life Tick*. Chicago: University of Chicago Press, 2016.

Rule, John C. "The Maurepas Papers: Portrait of a Minister." *French Historical Studies* 4, no. 1 (1965): 103–107.

Robinson, Dylan. *Hungry Listening: Resonant Theory for Indigenous Sound Studies*. Minneapolis: University of Minnesota Press, 2020.

Rodgers, Tara. *Pink Noises: Women on Electronic Music and Sound*. Durham, NC: Duke University Press, 2010.

Rodgers, Tara. "Tinkering with Cultural Memory: Gender and the Politics of Synthesizer Historiography." *Feminist Media Histories* 1, no. 4 (2015): 5–30.

Royster, Francesa T. *Sounding Like a No-No*. Ann Arbor: University of Michigan Press, 2013.

Rutledge, John. "How Did the Viola da Gamba Sound?" *Early Music* 7, no. 1 (1979): 59–69.

Safier, Niel. *Measuring the New World: Enlightenment Science and South America*. Chicago: University of Chicago Press, 2008.

Sanders, Wren. "Now List 2020: The Divine Mutability of Arca." *Them*, June 26, 2020. https://www.them.us/story/now-list-2020-arca-interview.

Sandry, Eleanor. "Creative Collaborations with Machines." *Philosophy of Technology* 30, no. 3 (2017): 305–319.

Savage, Roger. "Rameau's American Dancers." *Early Music* 11, no. 4 (1983): 441–452.

Savan, Jamie. "Revoicing a 'Choice Eunuch': The Cornett and Historical Models of Vocality." *Early Music* 46, no. 4 (2018): 561–578.

Sconce, Jeffrey. *Haunted Media: Electronic Presence from Telegraphy to Television*. Durham, NC: Duke University Press, 2000.

Seaver, Nick. *Computing Taste: Algorithms and the Makers of Music Recommendation*. Chicago: University of Chicago Press, 2022.

Shayt, David H. "Elephant under Glass: The Piano Key Bleach House of Deep River, Connecticut." *IA The Journal of the Society for Industrial Archeology* 19, no. 1 (1993): 37–59.

Showalter, English. *Françoise de Graffigny: Her Life and Works*. Oxford: Voltaire Foundation, 2004.

Siskin, Clifford, and William Warner, eds. *This Is Enlightenment*. Chicago: University of Chicago Press, 2010.

Smith, Stacy L., Marc Choueiti, Katherine Pieper. *Inclusion in the Recording Studio?* Annenberg Inclusion Initiative. Los Angeles: USC Annenberg School for Communication and Journalism, 2018.

Spary, Emma. *Utopia's Garden: French Natural History from Old Regime to Revolution*. Chicago: University of Chicago Press, 2000.

Spiers, Bradley M. "Music and the Spectacle of Artificial Life." PhD diss., University of Chicago, 2020.

Springer, Claudia. "Pleasure of the Interface." *Screen* 32, no. 3 (1991): 303–323.

Steingo, Gavin. "Whale Calling." *Ethnomusicology* 65, no. 2 (2021): 350–373.

Stoever, Jennifer Lynn. *The Sonic Color Line: Race and the Cultural Politics of Listening*. New York: NYU Press, 2016.

Suchman, Lucy. *Human-Machine Reconfigurations*. Cambridge, UK: Cambridge University Press, 2006.

Théberge, Paul. *Any Sound You Can Imagine: Making Music/Consuming Technology*. Middletown, CT: Wesleyan University Press, 1997.

Thomas, Downing A. *Music and the Origins of Language: Theories from the French Enlightenment*. Cambridge, UK: Cambridge University Press, 1995.
Thomas, Ruth P. "Françoise de'Issembourg de Graffigny (11 February 1695–21 December 1758)." In *Writers of the French Enlightenment I*, edited by Samia I. Spencer, vol. 313. Detroit: Gale, 2006.
Thurman, Kira. *Singing Like Germans: Black Musicians in the Land of Bach, Beethoven, and Brahms*. Ithaca, NY: Cornell University Press, 2021.
Tolley, Thomas. *Painting the Cannon's Roar: Music, the Visual Arts and the Rise of an Attentive Public in the Age of Haydn, c. 1750 to c. 1810*. New York: Routledge, 2001.
Trippet, David. "Music and the Transhuman Ear: Ultrasonics, Material Bodies, and the Limits of Sensation." *The Musical Quarterly* 100 (2018): 199–261.
Truitt, E. R. *Medieval Robots: Mechanism, Magic, Nature and Art*. Philadelphia: University of Pennsylvania Press, 2015.
Vágnerová, Lucie. "Sirens/Cyborgs: Sound Technologies and the Musical Body." PhD thesis, Columbia University, 2016.
Voskuhl, Adelheid. *Androids in the Enlightenment*. Chicago: University of Chicago Press, 2013.
Walker, Lance Scott. *DJ Screw: A Life in Slow Revolution*. Austin: University of Texas Press, 2022.
Weheliye, Alexander G. "'Feenin': Posthuman Voices in Contemporary Black Popular Music." *Social Text* 71 (2002): 21–47.
Westhoff, Ben. *Dirty South: Outkast, Lil Wayne, Soulja Boy, and the Southern Rappers Who Reinvented Hip-Hop*. Chicago: Chicago Review Press, 2011.
White, Sophie. *Wild Frenchmen and Frenchified Indians: Material Culture and Race in Colonial Louisiana*. Philadelphia: University of Pennsylvania Press, 2012.
Wright, Helena. *The Sex Factor in Marriage: A Book for Those Who Are About to Be Married*. New York: Vanguard, 1932.
Wynter, Sylvia. *The Hills of Hebron*. Kingston, Jamaica: Ian Randle Publishers, (1962) 2010.
Wynter, Sylvia. "Unsettling the Coloniality of Being/Power/Truth/Freedom: Towards the Human, After Man, Its Overrepresentation—An Argument." *CR: The New Centennial Review* 3, no. 3 (2003): 257–337.
Wynter, Sylvia, and Katherine McKittrick. "Unparalleled Catastrophe for Our Species? Or, to Give Humanness a Different Future: Conversations." In *Sylvia Wynter: On Being Human as Praxis*, edited by Katherine McKittrick, 9–89. Durham, NC: Duke University Press, 2015.
Yang, Mina. "East Meets West in the Concert Hall: Asians and Classical Music in the Century of Imperialism, Post Colonialism, and Multiculturalism." *Asian Music* 38, no. 1 (2007): 1–30.
Yearsley, David. *Bach and the Meanings of Counterpoint*. Cambridge, UK: Cambridge University Press, 2002.

Index

Page numbers followed by "f" refer to figures.

Abbate, Carolyn, 49, 51–52, 54, 115
Afrofuturism, 9–10, 141, 210n41
Allanbrook, Wye, 14, 53
analogies: human-machine, 12, 79–81, 137; human-musical instrument, 12–13, 32, 79–83, 110–11, 128. *See also under* artificial intelligence; Descartes, René; Diderot, Denis; Oram, Daphne
Anderson, Laurie, 144
androids: human-machine boundary, 9, 12, 21, 25, 49, 156–57, 176; and the uncanny, 7–9, 191n86. *See also* androids, musical; *Oracle, L'*
androids, musical / musical automata, 11–12, 15, 186n10; and anxiety, 19–21, 49, 136; and sensibility, 12, 19–21, 49, 136. *See also* de Caus, Salomon; *Oracle, L'*; Vaucanson flute player
Antares, 144, 146, 149
anthropomorphism, 13, 110–11, 113–16, 128, 159
Arca (Alejandra Ghersi), 173, 175; *Echo*, 171–72; "Riquiquí," 172–73, 174f
artificial intelligence (AI), 14, 79, 111; and analogies, 79–80, 161–63, 165, 167–72; and anxiety, 166–67, 167f, 170–71, 176; as autonomous creator, 159–61, 164–73, 175–76 (see also *Echo*; Howell, Emily; Spawn); and neural networks, 162–71; personification of (see *Echo*; Howell, Emily; Spawn); as tool, 160–61, 161f, 164–65
Auto-Tune, 13, 175, 211n64; and gender, 144–48; as posthuman, 13, 140–43, 176, 210n50; and purported musical deficiency, 138–39, 142; robotic vs. "human," 133, 138–40, 142, 146; and the vocoder, 139, 144–45

Bach, C. P. E., 124
Barzun, Jacques, 4–5, 79–80, 113, 115
Beethoven, Ludwig van: *Symphony no. 9* (1824) / "Ode to Joy," 10–11
bell, 63–64, 73–77, 176; and acoustics, 60–61; in bell tower (free swinging), 61, 68, 69f, 73; in carillon (instrument), 82–84; clock, 14–17, 76f, 77, 83–84, 86, 92, 105; mechanizing humans, 14–17, 51–52, 54; and timbre, 74–77; in viol music, 68, 69f, 73; and the voice, 52, 54, 75–77
Bennett, Jane, 115–16, 122, 131
Berland, Jody, 119–20
Blacking, John, 2–3
Blavet, Michel, 25, 45f, 187n26
body, human: and analogies (*see under* Descartes, René; Diderot, Denis); hands, 93, 96, 99–101, 101f, 175–76; and instrument playing, 43, 44–45f,

body, human (*cont.*)
    48, 56–58, 151–52; nerves/nervous system, 14, 80, 82–84, 87, 92, 107, 163–64; and the soul, 81–84, 90–91, 121
Bronze, 171–73, 174f

Carey, Mariah, 137–38, 142
Carlos, Wendy, 144, 211n55; Beethoven, "Ode to Joy," 10–11
Carr, CJ, 166–67
Cher, "Believe," 138, 144–45, 147, 150
Chopin, Frédéric, 1–2, 136, 181n6
Christensen, Thomas, 62–63, 194n34
Chua, Daniel, 53, 76
Chude Sokei, Louis, 9, 15, 71
Clark, Andy, 142, 156
Clayton, Jace, 142, 210n45
colonialism, 12, 41, 127, 140; and hearing/listening, 7, 63–64, 171; and hybrids/hybridity, 54, 65, 71–72; *Lettres d'une Péruvienne* (1747), 21, 46–48; and the *sauvage*, 39–41, 43, 46–48, 189–90n62; and slavery, 54, 71–72
Cope, David, 1–2, 159, 160f, 170
Couperin, François, 87–88, 88f, 89f
Coysevox, Antoine: *Faun Playing the Flute* (1709), 22–23, 23f, 24–25, 41
Cusick, Suzanne, 135, 149
cybernetics, 5, 137; and analogy, 79–80, 105
cyborgs, 52, 54, 134, 142, 144, 153, 156, 171; as hybrids, 137–38. *See also* Haraway, Donna
cyborg theory, 140–43, 156. *See also* Haraway, Donna
Cypess, Rebecca, 5, 110–11, 124

Dadabots, 14, 165–68, 167f, 171
Davies, Stephen, 116, 121
de Caus, Salomon, 56–57, 57f, 58f
DeepDream (Google), 162–63
dehumanization, 97, 135–36; as a positive, 67, 144; and vocal processing, 133–35, 144. *See also* posthumanism
de Machy, Le Sieur: *Pièces de Violle* (1685), 55–56, 58

Descartes, René, 79, 193n22; *Discours de la Méthode* (1637), 63, 188n43; human-musical instrument analogies, 79, 81–82, 86; and the soul, 82, 84, 126–27; *Traité de l'Homme / Treatise on Man* (ca. 1632/1664), 25–26, 31, 81–84
Dickinson, Kay, 144–45
Diderot, Denis: and gender, 89–92; human-clock analogy, 83–84, 86, 92, 105; human-harpsichord analogy, 13, 80–81, 86–92, 111, 123, 127, 172; human-vibrating strings analogy, 84–86; *Lettre sur les sourds et muets / Letter on the Deaf and Mute* (1751), 83–84, 92; *Le Neveu de Rameau / Rameau's Nephew*, 14, 53; *Le Rêve de d'Alembert / D'Alembert's Dream* (1769), 80–81, 84–86, 90, 92; and sensitivity, 84–86; and the soul, 84, 90–91, 127
Dodart, Denis, 73, 78
Dolan, Emily, 75, 183n26
Dryhurst, Mat, 168–70

Echo, 171–72
EDM (electronic dance music), 5–6, 134, 141, 143; and gender, 148–49, 153; and transhumanism, 143–44
Elkind, Rachel, 10–11
Enlightenment, the, 6, 9–10, 53, 92, 190n78; and automata, 12, 20, 136; human-machine boundary, 4, 12, 21, 49, 136; and machines, 6, 87; and the *sauvage*, 39–41, 43, 140
Eshun, Kodwo, 5, 9, 140–42, 156, 210n41

faun. *See* Coysevox, Antoine; mythology: and the faun; Vaucanson flute player
Ferrein, Antoine, 11; and hybrids/hybridity, 51–52, 74–75, 77–78; taxonomy of instruments, 12, 53–54, 74–75, 77–78
Festa, Lynn, 9, 21, 38, 49, 115
Fiebrink, Rebecca, 164
Flume: *Skin* (2016), 147

flute, transverse, 25–29, 35, 36–37f; left hand position of, 43, 44–45f, 48. See also mythology: Amphion; Triomphe des Arts, Le; Vaucanson flute player
Forqueray, Antoine, 68, 70, 73
Francoeur, François. See Piramé et Thisbe
Frere-Jones, Sasha, 138–40

gender: and electronic music, 99–101; euphoria, 173, 175; and musical instruments, 89–90, 151–52, 152f, 156, 159; and performance, 137–38, 151–56; and pop music production, 135, 144–50, 152–56
Gershwin, George, 96
Goehr, Lydia, 121–22
Graffigny, Françoise de, 11–12, 20–21; Lettres d'une Péruvienne (1747), 20–21, 46–48, 85; and L'Oracle, 29–30, 38; and sensibility, 19–21, 38, 46–49, 75, 81; and Vaucanson flute player, 12, 19–21, 29–30, 48–49
Gravelot, Hubert-François, 22f, 24, 41
Grimes (Claire Boucher), 145

Haraway, Donna, 4, 5, 52, 78, 137–38, 142, 143, 156, 171
harpsichord, 56, 87; and gender, 89–90; mottoes, 91–92, 123–24. See also Couperin, François
Harris, Calvin (Adam Wiles), 13; collab. with Rihanna, 133, 146–49, 152; and singing voice, 148–50, 155; "Summer," 150–51, 151f
Hayles, N. Katherine, 5, 81, 129–30, 155
Hegel, G. W. F.: Lectures on Aesthetics / Vorlesungen über die Ästhetik (1835), 121–22
Herndon, Holly, 14, 143–44, 147, 168–69
Hoffmann, E. T. A., 123, 191n86
Hotteterre, Jacques: Principes de la Flûte Traversière (1722), 25, 43, 44f
Houston, Whitney, 138, 154

Howell, Emily, 159–60, 160f, 170
Hu, Zhuqing (Lester), 63–64
human-machine boundary, 9, 12, 21, 25, 49, 51–52, 54, 80, 156–57, 176
hybrids/hybridity, 12, 51–54, 65–66, 70–75, 77–78. See also cyborgs; polyps; viol; voice

instruments, musical, 4, 54, 79–80, 175; as analogies, 12–13, 32, 79–83, 110–11, 128 (see also under Descartes, René; Diderot, Denis; Oram, Daphne); destruction of, 116–17 (see also piano, acoustic: as trash); gendering of, 89–90, 124, 151–52, 152f, 156, 159; as living, 123–24 (see also under piano, acoustic); as machines (see under bell; piano, acoustic); in myth (see mythology); partnership with, 124–26, 164–65, 170–71; and sampling, 129, 134–35, 153, 165; as soul prostheses, 116, 121–24, 131; subjugation of, 124–25, 151; taxonomies of, 12, 53–55, 66, 72, 74–78; and the voice, 54–56, 73–75, 155. See also Auto-Tune; bell; flute, transverse; harpsichord; lute; piano, acoustic; piano, non-acoustic; player piano; viol; violin; vocoder
Iyer, Vijay, 162, 169

Jlin, 168–70
Johnson, Barbara, 116, 128

Karagulla, Shafica, 107–8
Kesha, 139, 144
Kiiara (Kiara Saulters), 13, 133, 148, 153; collab. with Felix Snow, 146–47; "Gold," 133, 146–47, 153–54, 156
Kraftwerk, 5, 9–10
Kubrick, Stanley: 2001: A Space Odyssey, 1, 10–11, 139

La Barre, Michel de. See Triomphe des Arts, Le
Larsson, Zara, 13, 133; collab. with MNEK, 133, 146–47, 154–55; "Never Forget You," 133, 146, 154–55

Le Blanc, Hubert: *Defense de la basse de viole* (1740), 64–68, 70–73; and hybrids/hybridization, 12, 65–66, 70–73, 77–78; taxonomy of instruments, 53–54, 66, 72, 77–78

Lewis, George, 3, 214n7; *Voyager*, 161–62, 169–70

listening, 1–2, 15, 72, 130; to bells, 61, 73, 75, 77; and humanity, 4, 7, 12, 33, 52, 143, 153, 156–57; in instrument analogies, 82–84, 86; nighttime, 60–61, 73, 86. *See also* Couperin, François; Sauveur, Joseph

Liszt, Franz, 125, 151

Lockwood, Annea: "Piano Burning," 13, 117–18

lute, 55–57; in viol music, 68, 69f

Madonna, 144, 146

Mahaut, Antoine: *Nouvelle Méthode pour apprendre . . . a jouer de la flute traversière* (1759), 43, 44f

Marais, Marin, 59, 59f, 60, 64, 67–68, 69f, 73

Mascitti, Michele, 67–68, 70, 70f

McCarthy, John, 79–80

McLuhan, Marshall, 119–20, 141, 210n43

Mersenne, Marin: *Harmonie universelle* (1636), 54–56, 61, 74

Minogue, Kylie, 144, 146

MNEK (Uzoechi Emenike), 13, 133; collab. with Zara Larsson, 133, 146–47; "Never Forget You," 133, 146, 154–55; and singing voice, 154–55

Montagnat, Claude, 77–78

Mori, Masahiro, 8

Moseley, Roger, 91, 181n6, 183n26

Mozart, Wolfgang Amadeus, 5; *The Magic Flute*, 15, 51–52

mythology: Amphion, 26–28, 27f, 28f, 29f, 39; and the faun, 23–24; Orpheus, 56–57, 57f; *Piramé et Thisbe*, 35, 36–37f; Pygmalion, 26, 28–31, 30f, 47; and the *sauvage*, 39–41

neural networks. *See under* artificial intelligence

Ochoa Gautier, Ana María, 7, 135, 182n10

Offenbach, Jacques, *Les Contes d'Hoffmann*, 15

Oracle, L' (Saint-Foix, 1740), 29–33, 33f, 34–35, 52, 85, 188n36; and sensibility, 29–35, 38, 47–49

Oram, Daphne, 11, 92–93; compositional technique of, 93–100, 101f, 108–9; and the Computer Arts Society, 103, 104f, 105, 109–10; and the hand, 93, 96, 99–100, 101f; and human/machine analogies, 80–81, 93, 102–108, 111; *An Individual Note: Of Music, Sound and Electronics*, 11, 80–81, 103–4, 108–10; Oramics machine, 13, 80, 95–100, 101f, 108–11; and the paranormal, 105–6, 106f, 107–8; on serialism, 93–95, 100–101

organ, 56–58, 58f, 61; as analogy, 73–74, 81–83

Otis, Laura, 107, 197n3

Parreno, Philippe, 171–72

Pergolesi, Giovanni Battista: *La serva padrona*, 14–16

piano, acoustic: and anthropomorphism, 13, 115–16, 128; and consumerism, 115, 117–20, 130–31; immortality of, 118–21; keys, 127–30; as machine, 113, 119–22, 126, 131; and materiality, 117–21, 125–27, 129–30; personification of, 13, 113–14, 116, 121–26, 128, 131, 151–52; soul of, 122–23, 125–130; subjugation of, 124–25, 131; as trash, 13, 113–14, 114f, 115–18, 121–23, 128, 131. *See also* piano, non-acoustic; player piano

piano, non-acoustic, 13, 115, 128–30

Piramé et Thisbe (Francoeur, Rebel, La Serre, 1726/1740), 35, 36–37f

player piano, 119

polyps, 52–53, 90

popular music production, 134–35; and AI, 160–62, 165–67; and gender, 135, 144–56, 176; and posthumanism, 143–44, 155–57. *See also* Auto-Tune; sampling

posthumanism, 134–35, 141–44, 155–57; Black, 141. *See also* Afrofuturism
Powers, Ann, 142

Quantz, Johann Joachim, 2, 98, 176

race: and hybridity (*see under* colonialism); and machines, 9–10, 15; and piano keys, 127–28; and voice, 10, 13, 135, 138, 140, 154, 209n32. *See also* Afrofuturism; colonialism; posthumanism; Vaucanson flute player: as *homme sauvage*
Rameau, Jean-Philippe: and bells, 63–64, 76–77; *Dardanus* (1739), 48, 75; *Demonstration du principe de l'harmonie* (1750), 63; *Les Indes galantes* (1735), 39, 41, 189–90n62; *Nouveau système du musique theorique* (1726), 62, 65; *Observations sur notre instinct pour la musique* (1754), 63; *Pygmalion* (1748), 29; vs. J.-J. Rousseau, 63, 76–77; *Traité de l'Harmonie* (1722), 62; and the voice, 62–63, 65
Rebel, François. See *Piramé et Thisbe*
Reeps One, 167–68
Reynolds, Simon, 142, 210n50
Richards, Annette, 5, 204n38
Richards, Chris, 133–35, 143–44, 146–48, 155–56
Rihanna (Robyn Fenty), 13, 138, 150f; collab. with Calvin Harris, 133, 146–48, 152; "This Is What You Came For," 133, 146–48, 150, 150f, 152, 156
Riskin, Jessica, 7, 20, 84, 140, 186n10, 186n16, 198n5
Rose, Algernon, 125–26
Rosen, Jody, 142, 210n45
Rousseau, Jean, 55–56, 58–60, 64, 66, 75
Rousseau, Jean-Jacques: and bells, 76f, 76–77; vs. Rameau, 63, 76–77; and the voice, 52, 76–77, 85
Rubinstein, Anton, 125, 127

Saint-Foix, Germain-François Poullain de. See *Oracle, L'*
sampling, 133, 153–54; of instrument vs. voice, 134–35

sauvage. *See under* colonialism; mythology; Vaucanson flute player
Sauveur, Joseph, 60–63, 73
Sconce, Jeffrey, 107, 115
sensibility, 12, 26, 29–35, 38, 46–49, 171–72, 176
Shannon, Claude, 79–80
slavery, 9–10, 15, 54, 71–72, 141
Snow, Felix (William van der Heyden), 13, 133; collab. with Kiiara, 146–47, 153–54; and female voice, 149, 152–53
Sonami, Laetitia, 162, 163, 164–65, 169
Spawn, 168–69
Spears, Britney, 138–39, 142, 144
Stokowski, Leopold, 96
Streicher, Nannette, 12
Suchman, Lucy, 3–4
synthesizer, 2–3, 10–11, 134–35, 154–55, 161; and the vocoder, 9–11, 145. *See also* Oram, Daphne: Oramics machine

Tate, Greg, 141, 210n41
Taylor, Mark: "Believe" (Cher), 144–45, 147
"thing theory," 122–23. *See also* Bennett, Jane
T-Pain, 138–39, 142, 210n45
Trembley, Abraham, 52–53
Triomphe des Arts, Le (La Barre, La Motte, Pécour, 1700), 26–29, 27, 28f, 29f, 30f, 31, 35

Vaucanson, Jacques, 8; and the Académie Royale des Sciences, 22, 25; flute player automaton (*see* Vaucanson flute player); other automata, 39, 40f, 41–42
Vaucanson flute player, 11–12, 35, 38, 42f; as android, 12, 20; and anxiety, 8, 19–20, 49; construction of, 20–22, 22f, 24–25; and Coysevox statue, 21–25, 41, 43; as *homme sauvage*, 12, 39–43, 40f, 45–46, 48–49; and musical skill, 25, 136; and mythology, 12, 28–29; and pleasure, 8, 19–21, 25, 29, 49
viol: and bells, 68, 69f, 73; as hybrid, 12, 52–54, 65–66, 70–74; and the lute,

viol (*cont.*)
  55–57, 68, 69f; solo repertory, 55–56, 59–60, 66–69; technique, 57–59, 59f, 67–68, 72–73; vs. the violin, 12, 54, 64, 66–68, 70, 70f, 71–72, 78; and the voice, 54–56, 65–66, 70–71. *See also* Le Blanc, Hubert; Rousseau, Jean
violin, 109–10; solo repertory, 66–68, 70; vs. viol, 12, 54, 64, 66–68, 70, 70f, 71–72, 78
Vivares, François, 22f, 24, 41, 187n20
vocoder, 9–11, 133, 141, 144–45, 184n48, 184nn50–51, 211n55; vs. Auto-Tune, 139, 144–45
voice: and bells, 52, 54, 75–77; and harmony, 62–63; and humanity, 12, 51–52, 54–55, 73 (*see also* Auto-Tune; vocoder); as hybrid instrument, 12, 51–52, 77–78; and the viol, 54–56, 65–66, 70–71; as wind instrument, 73. *See also under* Ferrein, Antoine; Rameau, Jean-Philippe; Rousseau, Jean-Jacques
Voltaire (François- Marie Arouet), 20–21; *Alzire, ou les Américans*, 46
Voskuhl, Adelheid, 8, 21, 136

Walshe, Jennifer, 170–71
Weheliye, Alexander, 5–6, 10, 141, 208n13
Wekinator, 164
Wiener, Norbert, 79–80
Wonder, Stevie, 184n51
Wrench, Graham, 97–98
Wynter, Sylvia, 135, 156, 208n13, 217n49

Zinovieff, Peter, 99–101
Zukowski, Zach (Dadabots), 165–66

www.ingramcontent.com/pod-product-compliance
Lightning Source LLC
Chambersburg PA
CBHW022049290426
44109CB00014B/1037